SCIENCE AND MORAL PRIORITY

CONVERGENCE

Founded, Planned, and Edited by
RUTH NANDA ANSHEN

Board of Editors
of
CONVERGENCE

SCIENCE AND MORAL PRIORITY

ROGER SPERRY

Merging Mind,
Brain, and Human Values

1983
COLUMBIA UNIVERSITY PRESS
NEW YORK

Library of Congress Cataloging in Publication Data

Sperry, Roger Wolcott, 1913–
Science and moral priority.

(Convergence)
Bibliograpy: p.
1. Neuropsychology—Philosophy.
2. Ethics.   3. Intellect.   4. Science—
Philosophy.   I. Title.   II. Series:
Convergence (New York, N.Y.)
QP360.S63   174'.95   81-24206
ISBN 0-231-05406-8   AACR2

Columbia University Press
New York   Guildford, Surrey

Clothbound editions of Columbia University Press books are Smyth-
sewn and printed on permanent and durable acid-free paper.

—TO THOSE MANY
GENERATIONS
HOPEFULLY TO COME

Acknowledgments

I am grateful to Dr. Ruth Nanda Anshen for suggesting this volume and for her editorial help in our efforts to mold these collected essays into a single unified volume adapted in style for the general informed reader. We thank the publishers of the original material for permission to use it in the present form and context. Prior journal and book sources are listed in the *References* and *Notes* sections. The original work and research referred to has been generously supported over the years by the California Institute of Technology through the Frank P. Hixon Fund and by grants from the National Institute of Mental Health. Special projects were aided also by the National Science Foundation, the David Stone Foundation, and the Pew Memorial Trust Fund. I am grateful to Erika Erdmann for help with library research and to Lois MacBird for assistance in assembling the manuscript material and for library, laboratory, and related aid extending over many years during work on the original material. I thank the numerous colleagues, editors, referees and others who have made constructive suggestions on earlier drafts of the manuscript—and lasting thanks to my wife, Norma Gay, for valuable feedback, personal support, and patient effort throughout to create favorable conditions for the work.

Contents

Foreword

Because he seeks a particular truth in objective experience, the scientist rarely makes a contribution to the philosophy of morals. He hoes the fields of knowledge without looking far into the landscape or deep within himself. Trying to keep his mind on the facts and task at hand, he resists involvement in loose speculation or passionate argument. The successful scientist must have disciplined thought and be dedicated to unbiased investigation agreeing always to conditions which can be identified and measured in terms that are universally acceptable. The nature of creative inquiry draws his mental focus into such a narrow channel that he may tend in time to know much less than his neighbors about human affairs, about beliefs and perspectives regarding life as a whole, and especially about the irrational sources of interpersonal life.

And yet, from time to time it is a scientist who changes the way all of us perceive ourselves and each other, as well as the way we look at the world. Thus the young Charles Darwin's reflections on nature's patterns forced him to challenge some of the deepest values of men he admired and of family near and dear. Some of Darwin's researches after he became a devoted naturalist were extremely specialized, even pedantic in detail; an encyclopedic work on barnacles drove him to desolation. Yet, detecting an undeciphered message in nature, he continued his examinations on orchids and honey bees, volcanoes and fossils, exploring and struggling for half a century to interpret this message—until he became the most discussed thinker of his age.

In its highest form, the scientific belief in nature may even lead to a spiritual insight that can stand against bigotry and superstition in established dogma. Einstein wrote of a cosmic religious sense that comes from mystic revelation and from the most advanced contemplation of the intricate order in nature. He claimed that this cosmic awareness might become a higher, more developed form of religion which would increase the sense of meaning, without supplanting pantheistic appeasement of nature's threats or human-centered religious codes that more commonly comfort loneliness and the anguish of adversity.

Roger Sperry's scientific life, directed largely to mysteries within man's inner being, has been guided by a persistent quest for understanding of one of nature's greatest riddles—the relation of mind to brain. Like Darwin and Einstein, he is led through his pursuit of science to a changed view of the world and to a religious philosophy in which the cosmic order of evolving nature is seen to transcend, though not exclude, the more immediate personal values and needs of mankind. He sees current mind-brain science upholding a framework for moral values in which the human psyche, though the prime, crowning determinant of nature, is not the final measure of all things. His argument demands that more "godlike perspectives," referent to all creation, be placed above otherwise compelling humanitarian rules of conduct when the two appear in conflict.

The contributions of Sperry to the conceptualization of mind and brain require change in the basic philosophy of science itself. Traditional mechanistic determinism of twentieth-century science is replaced by a new philosophy in which nature's highest and most evolved manifestations attain causal control over the fate of entities at lower levels. This concept of a causal potency in consciousness capable of holding influence downward over all component orders of cerebral function, with the linked theory that values inherent in the patterns of the mind are the key determinants in all decision-making and a natural topic for scientific as well as philosophical enquiry, form the heart of his argument.

Sperry's philosophical message is firmly rooted in a masterly knowledge of the life of the brain starting with its growth in the embryo. In early experiments on the plasticity and developmental specification of brain circuits that gained him fame before he was forty, Sperry showed step by step that every main type of linkage in the brain could have its significant structure determined from within the developmental process itself. He concluded that the patterns of experience, or the layout of environmental stimulation, were insignificant for the fundamental plan of this development. The immediate environment was not the sole informant of the more complex psychological functions, as had been assumed in the reflex physiology of Pavlov and in Watson's behaviorism. The surgical and experimental techniques and the strategy of interventions in Sperry's work on formation of nerve circuits were brilliant and established him in a unique lead position among the then small band of researchers who sought a biological theory of psychological function.

It was his work on the surgically divided brain that led Sperry to a direct confrontation with the creative force of consciousness. Begun in the early 1950s with Ronald Myers in Chicago, and continued with a succession of graduate students and other associates at the California Institute of Technology, psychological studies on split-brain animals led the way to startling discoveries on human beings who had undergone similar surgery for control of intractable epilepsy. The findings, concerning a strangely divided mental state in which two different consciousnesses may cohabit the same skull in harmony, have profoundly interested philosophers and opened a great new territory of inquiry. They have brought psychologists and neurologists to consider more closely than before the relationship between functions of the mind and of the brain, and to ponder the anatomy of the "self." Although the thinking behind Sperry's theory of consciousness as a causal force in brain activity had a considerable history, it was these studies on the surgically divided mind that prompted the first developed expression and publication of the new philosophy in the mid-sixties.

The issues of consciousness, brain, and moral values have since become increasingly the subject of hot debate in professional journals of philosophy, psychology, neuroscience, and even religion. Few brain scientists, however, have attempted to counter

Sperry's proposals, which after all imply a major shift in neuro-biological thinking. Although his mind-brain concepts have been used to bolster old dualist arguments, on the one side, and materialist mind-brain identity philosophy on the other, Sperry prefers to think of his position as neither one of these, but more correctly described as a distinct, intermediate framework of thought. Under labels such as "mentalist monism," or "emergent interactionism," it offers the one coherent theory of self-regulated motives for consciously controlled actions, and it is built of concepts that flow directly from research on the brain. For this reason, if for no other, these ideas are unique. The novel insights of this volume promise to guide both speculative philosophy and scientific examination of the evidence for years to come in the search for a new foundation of belief. This search must acknowledge the inherent wisdom and power of the human mind and its necessary fitness to the nature of experience. If Sperry is right, the psycho-biological approach will create values to guide humanity to a higher path of survival than that uncertain course we presently follow.

Colwyn Trevarthen
Department of Psychology
University of Edinburgh
Scotland

SCIENCE AND MORAL PRIORITY

For I have learned
To look on nature, not as in the hour
Of thoughtless youth; but hearing oftentimes
The still, sad music of humanity,
Nor harsh nor grating, though of ample power
To chasten and subdue.

—Wordsworth 1798

Introduction

Timeless Riddle of Right and Wrong

We are concerned here in large part with the problem of moral choice: the question of what is right and what is wrong. This is something with which mankind has grappled since the dawn of human conscience and thus far failed to reach any satisfying solution. Bernard Shaw, writing in *Major Barbara* describes it as:

> **the secret that has puzzled all the philosophers, baffled all the lawyers, muddled all the men of business, and ruined most of the artists: the secret of right and wrong.**

Opposing views derive from differences in the diverse and conflicting faiths and value systems of various peoples and can hardly be resolved unless these latter are brought into harmony.

When ethical and moral values stand in conflict, how do we decide which alternative to uphold and by what ultimate standards and criteria? The search for answers leads quickly into issues inseparable from the historically impassioned controversy over what is most sacred. To a large extent the quest for the secret to right and wrong becomes a search for the "highest good" or "ultimate value." This in turn becomes inextricably interfused with ageless concerns about man's role in the scheme of things and the meaning of existence.

One may ask why this seemingly insoluble question that has baffled analysis and explanation for centuries should again be reopened. The reasons go back to recent advances in concepts of mind and brain and related progress in our understanding of the origins and structure of human value systems: developments that have changed the foundational framework for attacking the problem, opened new perspectives on the interrelation of science and values and bring, overall, a new level of comprehension that calls for new approaches with possibilities for new answers.

An important outcome from the standpoint of relevance is the suggested remedy that emerges for some of the grave global problems of our times. Rather than propose new sources of energy or ways to produce more food, curb pollution, restore our cities and the like, the suggested treatment for global ills is aimed more directly at the source and underlying causes. Starting from a recent editorial in the journal *Science,* we can begin with the obvious, but rarely publicized, observation that "It is people who use energy. With fewer people we would need less energy." It is also people who consume food, overcrowd cities, create pollution, wreak ecological havoc, and all the rest. Unless population growth is controlled, other remedial measures must clearly fail in the long run. Yet, the mere thought of controlling the multiplication of people immediately runs head-on into conflict with all kinds of sensitive, centuries-old ethical and moral conventions, raising a host of entangled pros and cons in human value issues.

It is these touchy human value, moral choice factors to which the present writings are directed and which must first be faced or somehow dealt with before any global action will be effective. Initially my concern with value theory came about only secondarily as a spin-off of work on mind-brain relations, but very shortly, as the key importance and many humanistic implications of my work became apparent, the tail began to wag the dog. I was led into an involved critical examination of the relation of science to values and of the basic foundations and structure of value-belief systems, questioning the implicit premises, fundamental criteria, and the ultimate frame of reference behind ethical and moral standards—all from the standpoint of the recent new views in the mind-brain and neurobehavioral sciences. The outcome is my strong conviction that these emotionally charged, but highly critical, human value issues are best faced and dealt with openly in rational terms.

Do Moral Values Transcend Reason?

This is in spite of the fact that values, with their interwoven questions of moral right and wrong, have always been notoriously resistant to any rational treatment or understanding. The endless complexity of human values and their subjectivity, their relativity to changing aims and goals, and their frequent irrationality and anchorage in intuition and religious faith, among other difficulties, have seemed to hopelessly defy any rational or scientific approach. Both science and philosophy have long taught that no proof for any of our most prized values can ever be demonstrated by the scientific method. It is claimed that the same set of scientific data can be used to support directly opposed values, that it is logically impossible to derive subjective values from objective facts, or to logically infer what ethically *ought* to be, from a description of what actually *is*.

Reasoning of this kind, which has served to keep science and values separate in the past, is today widely undermined. New concepts of mind and brain have helped to bring a profound change of pardigm in behavioral sciences known as the "consciousness" or "cognitive" revolution, also referred to in some quarters as the "humanist" or "third" revolution. Based on the shift to a new causal interpretation of conscious experience, it has overthrown long-accepted ideas of twentieth-century scientific materialism. The repercussions of the new view of consciousness pervade the entire structure of the scientific approach to human nature, affecting in multiple ways the rational treatment of moral values. Emergent new views stress the primacy of inner experience, revise interpretations of the conscious self, and offer new insights into the nature of values, freedom of choice, personal transcendence, and afterlife possibilities. The implications extend further into basic underpinnings of science, modifying its world view and its description of physical reality. Although the changes in many cases are still in progress, or too recent to permit a full appraisal, the collective consequences clearly promise to profoundly affect our values and touch our lives in many ways. Already they are seen as altering the scope, world view, and humanist role of science, and revising the ultimate criteria for value and meaning.

Turnabout in Scientific Status of Consciousness

Until very recently science had been dominated in the Western and Communist worlds alike by the belief that man and his behavior, along with everything else, can be fully accounted for in terms that are strictly material without resorting to any kind of nonphysical force or agent. Among the latter were traditionally included all the intangibles of mind and spirit: mental images, sensations, thoughts and feelings, hopes, ideals, and all the other subjective phenomena that comprise the world of inner experience. So strong had been the dogmatic renunciation of conscious or mental forces as explanatory constructs in natural science up through the 1950s and into the early 1960s that one risked derision by even mentioning words such as "conscious" or "mental" at a serious scientific gathering.

Our new acceptance of conscious entities as causal, that began in the latter 1960s and literally exploded in the 1970s, gives a new look to science and what science stands for. On the new terms that include recognition of emergent or "downward" causation, we no longer believe everything, including the human psyche, is reducible in principle to quantum mechanics. The forces of mind and consciousness are perceived to supersede those of biophysics, chemistry, and physiology. Whereas on the old terms, science tended by nature to be dehumanizing, destroying value and meaning, and had to be kept separate from value judgment, the way is now opened for a congenial mergence of science with the value disciplines. The basic conflict of worldview between science and the humanities becomes resolved, while philosophy, in a switch from its late preoccupation with language, once again finds values and "the good life" to be worthy subjects.

A revitalized approach to value problems now becomes possible, and leads to a recurring conclusion, reinforced from many different angles, that the prime hope for tomorrow's world lies not in outer space or improved technology, but rather in a change in the kinds of value-belief systems we live and govern by. Essentially the same conclusion was drawn at a meeting sponsored by the National Council of Churches in Washington, D.C., 1980, where it was unanimously agreed—by Protestant, Catholic, Jewish and other faiths—that what the world needs is a new theology, one that would promote the values of conservation, renewable energy sources, and the like. Values of this kind appear to be exactly what would emerge from a union of science, on our new terms, with ethics and religion.

Human Values Shape History

It should come as no great surprise that an in-depth treatment of value systems should lead to logistics for world change. The tremendous power of human values to mold world conditions and decide the course of history can hardly be overstated. Human values shape human decisions that in turn govern human destiny. Any handle on human values becomes a potential handle on the future. It follows that to turn the menacing tide of spiraling population and corollary disaster trends, we need not wait for a nuclear holocaust, global famine, massive irreversible destruction of species, life-support systems, and the like. A mere shift in human values will do it. All it would require is a relatively painless adjustment in mankind's sense of right and wrong.

Much of the present argument is devoted to showing that a synthesis of science with moral values, despite previous views to the contrary, is today logically feasible, humanistically compatible and scientifically sound. To what extent it may also be acceptable or desirable from the religious standpoint remains to be seen. Proposals to unite religion and science have been with us for a long time without gaining any appreciable popular support. One would hardly have the temerity to hope that the decades-old status of this effort might be altered, were it not, again, for the recent changes within mind-brain science that have fundamentally transformed what science contributes to such a union.

Beginning with what can be accepted axiomatically as given, inherent, or self-evident, and building on the kind of reality upheld by science, we come out with the outlines of a value-belief system that would appear to be something that might stand a chance at the United Nations. Prospects for world government have remained dim in large part because peoples of differing faiths, cultures, and ideologies are loath to submit to being ruled by the values of opposing belief systems for which they have no credence or empathy and often little respect. One can see some glimmer of hope, however, that Christians, Communists, capitalists, Hindus, Buddhists, and all the others might be willing to agree on, and respect, the kinds of values and beliefs founded in the validity and worldview of present-day nonreductive science—at least for purposes of compromise and world government.

The present collection of essays, spanning a sixteen-year period of philosophic transition, takes off at a time when human values, from the standpoint of science, remained a bewildering confusion set apart in an objectively unapproachable subjective realm. Even in philosophy there was little interest at the time in the problem of values. My own efforts to follow the human value implications of our science into this unpromising realm, despite the shortcomings and my lack of background, were impelled by what seemed the overwhelmingly urgent importance of bringing to general awareness an understanding of these new developments and what they mean, regardless of how crude or polished the effort.

Chapter 1 challenges the traditional "beyond science" interpretation of values in a first attempt to apply our new mind-brain notions. In chapter 2 we confront what W. T. Jones (34) calls "the crisis of contemporary culture," namely, the profound contradiction between the traditional humanist views of man and the world on the one hand and the value devoid, mechanistic descriptions of science on the other, a disparity largely responsible for the chasm of culture between scientists and humanists described by C. P. Snow (66). A unifying resolution is proposed via corrective changes in the interpretations of science. These in turn open the way for a genuine fusion of science and values, various aspects and consequences of which are explored in the remaining chapters. Chapter 9 comes closest to an up-to-date summary of the message of the volume.

One quickly learns on venturing into the arena of human values that there is no more difficult subject on which to write without somehow offending somebody's sensibilities, including one's own (as I look back years later on earlier writings). However, even though a more appealing and sophisticated statement might be possible today, there is something to be said for starting from scratch and retracing the arguments through the early formative phases in the original context.

The reader will find that the different approaches and ideas of the separate chapters all contribute to and reinforce a central thesis that should build consistently toward a coherent new outlook. For purposes of the volume as a whole, some rearrangement has been made in the original order, titles revised, repetitions deleted, subheads added, and some minor editorial revisions made to obtain a more natural progression for general reading. I hope that with a subject of this kind many readers will find the remaining overlap welcome and helpful.

Roger Sperry
Pasadena, California
September 1981

1

Values: Number One Problem of Our Times

And I have felt
A presence that disturbs me with the joy
Of elevated thoughts; a sense sublime
Of something far more deeply interfused,
Whose dwelling is the light of setting suns,
And the round ocean and the living air, . . .
—Wordsworth 1798

By evolutionary time standards, the fate of life on our planet has suddenly and quite abruptly come to rest on an entirely new form of security and control, based on the machinery of the human brain. The older, noncognitive controls of nature that have regulated events in our biosphere for hundreds of millions of years, the forces of nature that lifted life from the protozoan to the human level and created man, are no longer in command. Modern man has intervened and now superimposes on nature his own cognitive brand of global domination. The outstanding feature of our times is the occurrence of this radical shift in biospheric controls away from the vast interwoven matrix of pluralistic, time-tested checks and balances of nature, to the much more arbitrary, monistic, and relatively untested mental capacities and impulses of the human brain.

World's Most Potent Force

Along with its weaknesses, our newly imposed human system of global regulation also contains tremendous new powers, including the potential to effect changes within a decade that formerly required thousands and millions of years. Almost the entire fabric of the earth's surface, from the atomic to the scenic level, is rapidly becoming subject to disassembly and resynthesis along new patterns of human design. In all this human-directed supervision, the potential for utopian advancement throughout the globe seems endless. It is important that these utopian potentialities be recognized and remembered as we turn now to consider the other side of the coin.

Despite the beneficial features of human domination, it becomes increasingly apparent that our biosphere is set today on a disaster course as a direct consequence of human intervention. The entire grand design of life, painstakingly evolved over millennia, suddenly is subject to instant destruction, depending only on some passing twist in human affairs. If nuclear extermination is avoided, other built-in, self-destruct features are evident that threaten to bring all civilization to a halt—if things continue as they are going (17, 24).

Some modern analysts are inclined to put the blame for the mounting world crises primarily on excessive population; others blame science and technology; some point to creeping materialism and the pursuit of economic gain, and to the loss of faith and of moral values; Communists accuse capitalism, and vice versa; some emphasize racism and intolerance, while others deplore dysgenic trends in the population. Although the apparent causes are multiple, complex, and confusing at the political, economic, and social levels of analysis, a common root source of dysfunction can be seen when the situation is viewed more objectively through the broad perspectives of evolution and the life and behavioral sciences. In short, if we could summon an extraterrestrial troubleshooter to examine our earthly predicament with an outer space perspective free of human bias, I believe he very quickly would put his finger on the human value factor in our biospheric controls as the primary underlying cause of most of our difficulties.

In other words, his examination would show that the trends toward disaster in today's world stem mainly from the fact that while man has been acquiring new, almost godlike, powers of control over nature, he has continued to wield these same powers with a relatively shortsighted, most ungodlike set of values, rooted, on the one hand, in outdated biologic hangovers from evolution in the Stone Age and, on the other, in various mythologies and ideologies based on little more than faith, fantasy, wishful thinking, altered mental states, and intuition. The obvious recommendation is to shape up our value systems to something more in tune with present-day reality, more properly suited to the new powers that man now commands and the new problems we now face. It might be added that any attempt to attack directly the overt symptoms of our global condition—pollution, poverty, aggression, overpopulation, and so on—can hardly succeed until the requisite changes are first achieved in the underlying human values involved. Once the subjective value factor has been adjusted, corrections will follow readily in the more concrete features of the system.

The reasoning behind these blunt statements is more lengthy. At the outset let it be taken for granted that a reciprocal causal interaction exists between values and related technological, economic, and social conditions. Our subjective values, that is, not only *reflect* environmental conditions but also *produce and control* them. Any complex cycle, spiral, or four-dimensional latticework of causal interactions, like that involving human values and environmental conditions, can be interrupted and shaped from numerous points in the system. Why, then, the selective focus on the value factor? Why single out this particular feature of the total causal complex as the one where corrective change is most needed, and the one where remedial effort would be most strategically directed? The answers are complex and call for an objective understanding of the basis of human values, their origins and structure, and particularly for a more widespread recognition of the critical role that values play as causal agents in the biospheric chain of control.

The human brain is today the dominant control force on our planet; what moves and directs the brain of man will, in turn, largely determine the future. Among that vast complex of forces that influence and control the brain and behavior of man, the factor of human values stands out as a universal determinant of all human decisions and action. Every voluntary act and/or decision by an individual or a group inevitably is governed, overtly or implicitly, by value priorities. In essence, what a person or a society values determines what it does. The human value factor, defined in this way and viewed objectively in terms of brain states that govern acts, thoughts, and decisions, may be seen to occupy a central position of strategic regulative influence in the total biospheric chain of command.

The Underlying Master Control

One can agree with those who claim that excess population is the principal potentiating factor behind a large majority of today's problems. Yet, behind the population surplus one sees always the determining factor of human values with which it is necessary to cope first to attain any effective control over human procreation. The same reasoning will be found to apply to other major threats like pollution, poverty, war, and nuclear escalation.

What man does to his world will be determined very largely by the subjective values and beliefs by which he lives and is moved and guided. As human numbers increase, and as science and technology grow ever more powerful, the greater becomes the strategic control power of the human value factor that determines how all of this growing human power will be applied and directed. Simple logic says that future alterations in this single factor alone could spell the difference between utopia and social disaster. Viewed objectively as top-level causal agents in our global control system, human values have become too important to be treated, as in the past, simply by neglect or by a laissez-faire or even a "hands-off" policy. The new conditions call for a new concern and a new approach.

The current widespread rejection and breakdown of the mainline value and belief systems by which civilized man has lived for centuries have additionally amplified in recent years the need for constructive adjustment in the value factor as such. While the "God is dead" and related movements of the past decade have resulted in considerable searching for and testing of new values and new life styles, these groping efforts have not yet succeeded in replacing the old discarded guidelines with new ones, at least not on any scale sufficient to be socially effective. With this gap unfilled, large segments of civilized society drift today in a state of confusion, at a loss with regard to ethical standards, morality, goals, and a sense of purpose and direction in the human endeavor generally.

When the Society for Zero Population Growth squares off against the Church on issues of abortion, birth control, optimal population, and related questions, by what ultimate standards do we decide who is in the right? Similarly, when other opposing factions come to fundamental philosophic disagreement on issues like justifiable military killing, human exploitation of other species, eugenics, euthanasia, plunder of natural resources, noble savagery versus the urban rat race, redwoods versus freeways, and the multitude of other value questions that now confront us, by what ultimates do we attempt to distinguish right from wrong? Our tolerant, educated Western societies, in particular, seem more and more to be lacking in conviction with regard to any kind of ultimate standards.

Societal values tend to be self-corrective to a large degree, and to change naturally in response to changing needs and conditions, but in these days of extremely rapid change the time lag is defeative. By the time a voting majority becomes ready to recognize and endorse new values, as it now seems to be doing with respect to pollution and overpopulation, the situation will already have advanced far beyond the state of the optimal ideal toward a condition of intolerability. As long as values are formed on this feedback basis, social existence will continue to fluctuate around levels of survival and tolerability rather than those of any ideal. Wherever possible, it is therefore preferable that value developments precede and help control, rather than follow, changes in social conditions. In this same connection there is a good basis for concluding that the human brain, with its advanced cognitive capabilities, does better to seek its values above the natural, immediate, situational level, in more rational, long-term, and idealistic realms.

For these and reasons to follow, it seems important that the social value factor be more generally recognized as a powerful causal agent in its own right, and something to be dealt with directly as such. No more critical task can be projected for the future than that of seeking for civilized society a new, elevated set of value guidelines more suited to man's expanding numbers and new powers over nature, a frame of reference for value priorities that will act to secure and conserve our world instead of destroying it. My suggestion here is that answers can be found through a fusion of science, ethics, and religion that would bring the truths, insights, worldview reliability and other attributes and benefits of science to bear upon the whole problem of values and value priorities. There is need, as a starting basis, for what might almost be called a science of values. This, with what follows, is the personal conviction that emerges from my experience and perspectives in the mind-brain and related life sciences. No claim can be made for its originality or sophistication. Mainly, it constitutes an explanatory enlargement and defense of an earlier contention regarding the feasibility of merging science and values (74).

Traditional Views Obsolete

At first thought values appear to be entirely impossible to treat on any rational, logical, or scientific basis. Human values, as a reflection of man's beliefs, wants, needs, and ideology as well as of more concrete biological and situational conditions are subjective and often irrational. More than this, the basic values of a people are tied in closely with their religious beliefs and with "inalienable" personal and civil rights, freedom, and the like. In the minds of most of us, human values acquire a kind of inviolate sacred primacy which makes them immune to any deliberate analysis and corrective alteration for other ends. Thus, the human value factor has been not only neglected or treated only indirectly, but often pointedly bypassed by policy in efforts to remedy global conditions.

Resistance to a union of science and values stems further from a traditional understanding that value questions, by nature, are beyond science. "Value judgments lie outside the realm of science," we are told, and, "Science may tell us *how*, but not *why*," or "Science can tell us how to achieve a given goal but cannot tell which goals to aim for." Thus, on the one hand, we have human values as the paramount problem of our time; on the other, we have science as the proven "Number 1" method available for answering problems, and obtaining the kind of validity on which values should be based. Paradoxically, we are taught that the two belong in separate realms and that the one must not be applied to the other.

Unable, in the light of our current mind-brain theory, to accept any longer this traditional separation of science and values, I will argue that not only are science and values quite miscible and a scientific approach both feasible and desirable but further, that the best foundation and reference frame for moral values can be found in the kind of validity and this-world reality supported by science on the new present-day terms to be explained shortly. This derives in part from a conviction that the functional organization of our cerebral machinery is intrinsically such that the scientific method offers the most reliable means by which a brain can arrive at an operationally valid stance in the realm of values, as well as elsewhere.

The desirability of attempting to bring constraints of science to apply to the problem of values will continue to require justification on many counts. Formal religious doctrine historically has set the foundations for civilized man's ultimate values—the top values, that is, that are the final referents behind the systems of subsidiary values that comprise the daily "good life" of whatever one's faith may be. The mere thought of exposing these spiritual and sacred ultimates to the open scrutiny, analysis, and manipulative empiricism of science may cause a shudder in many quarters. I will try to show below that fears along these lines can be largely dispelled.

Through progressive undermining of sacred dogma, science long has been regarded more as archenemy than ally of religion and religious values. In addition, society seems increasingly inclined these days to look to the spirit of antiscience for solutions. Science and technology stand accused of having created many of the crisis problems that now confront us. For our present argument it is important to recognize that science is being blamed here not because it has failed, but, on the contrary, because it has succeeded so well. What has failed is not science but rather the value and belief systems that have determined the way in which scientific advances have been applied.

Science and the scientific method are supposed to involve objective measurements of cold, value free, quantitative phenomena, and therefore to be inherently disqualified to deal with subjective values. This argument may have had some philosophic validity in the past, particularly with reference to the physical sciences, but it fails to take into account the content, principles, and phenomena of the behavioral and life sciences as developed today. Modern behavioral science deals directly with value preferences and their formation as important causal variables in behavior, and it also deals with goals, needs, motivation, and related factors at individual, group, and social levels. The origin, development, and causal role of values are now very much a part of science.

A related argument would keep science and values separate by asserting that values are subjective mental phenomena and thus inaccessible to objective science. This dualistic logic also is no longer applicable. Current theory of mind leads to a quite different philosophy regarding the relation of objective factual science to subjective experience. Mental awareness no longer need be set off in separate metaphysical, epiphenomenal, or other parallelistic or dualistic realms (74, 77). Subjective values, like other mental phenomena, become an integral part of the objective brain process with top-level control potency in the sequence of causation in man's decision-making machinery. In these current terms, subjective values can be treated in principle, along with facts, as causal agents in brain processing and in the objective world, and thus are a legitimate concern of objective science.

Another old argument holds that the scope of science is inadequate to be of much help with questions concerning the ultimate goals and meaning of existence with which religion deals, and which largely set the basic parameters for social values. This gap between religion and science has been largely erased by modern advances in our concepts of cosmology, the nature of matter, the forces that move the universe and created life, and the nature of mind and the mind-brain relation. All these advanced insights make science today highly relevant and directly competitive in scope with revelation, faith, and intuition.

Clearly, many of the traditional reasons for discounting a scientific approach to values do not hold up under examination today. Certain aspects of values, however, will continue to pose difficulties for science. On examination these remaining difficulties are found to apply as well to the alternative sources of values like intuition, common sense, and faith, as well as to political, legal, or economic philosophy. Regardless of all the various difficulties, society does have to get its values from somewhere, and at present it seems a fair statement that man has no guidelines for obtaining social values that are superior to those set by the validity and world view of science. More than intuition—and just as much as revelation, politics, law, and other disciplines, including philosophy and religion—science deals with ultimates.

Toward a Hierarchic Theory of Values

If objections in principle can be removed and the way formally cleared for an open, rational approach in the realm of values, the question remains as to whether any significant practical benefit can be expected. Perhaps human values are so enormously complex, amorphous, irrational, relative, and generally intangible that any attempt at a scientific approach becomes hopelessly entangled from the start? To the scientist who likes to see order instead of chaos, certainty instead of myth, who wants to create a systematized body of knowledge and to understand causal and logical interrelationships, and perhaps do something about predicting and controlling the consequences, the field of human values certainly presents a formidable challenge.

Some fundamental points about the nature and origins of values are discernible, however, and do much to help prepare the way. First, values of the cognitive, ideological sort, that are of the greatest concern for our present purposes, are found to exist in relationship to directional or aimed activity in hierarchical systems and subsystems that are goal-dependent. Given any desired goal, that which helps toward attainment of the goal becomes good, and that which obstructs the goal becomes bad. Similarly, everything that helps to attain all the subsidiary aims which in turn help to reach the main goal, also becomes valued accordingly. A shift in the main goal may bring corresponding shifts—even reversals of value—throughout the whole associated hierarchy of subsidiary values.

It follows, further, that any concept or belief that is accepted regarding the goal and value of life as a whole will then logically supersede and determine values at subsidiary levels. This is why religion, and philosophy to a lesser degree, by postulating answers at the top levels thereby become a final authority for value judgments in general. Legal, moral, and other codes must conform, and, in case of conflict, deference is commonly given to a person's religious conscience and convictions. What is sacred gets special priority. Once a given answer regarding the goal or meaning of existence becomes accepted, value priorities then can be ordered and value issues judged accordingly. Whether the aim be a place in heaven, a suspended state of nirvana, progress of the "party," or whatever, the good life and the converse automatically crystallize by logical inference from any accepted belief concerning the ultimate goal.

The focus here and throughout this volume is on cognitive values at the ideological level, because these are the values that are of major importance in the global chain of command. In the long-term, large-scale sociopolitical activities, cultural conflicts, and ideological power struggles that are of concern in today's crisis problems, these cognitive values outweigh and tend to supersede the more immediate, situational, irrational, and natural or biological values.

Social values are necessarily built in large part upon and around inherent value traits in human nature written into the species by evolution (25, 29). These basic species traits were excellent for assisting evolution and survival through the Stone Age, but become disastrous when combined with the overwhelming numbers and technological powers of modern man. The basic aspects of human nature and their direct value derivatives are treated here as constants in the total picture, to be accepted and worked with, rather than modified. Fortunately, the social consequences of values of this kind are subject to considerable regulation and control through the higher cognitive value systems reinforced by cultural codes and written law. The value of staying out of prison may be utilized to control undesirable natural impulses. It is the manmade laws, written and unwritten, enforcing values of cognitive origin, about which one can hope to do something. Thus the large "human nature" element in the value problem, and along with it much of the "irrational" aspect, is taken care of, if one can properly manage the supersedent systems of ideological values.

Ultimate Foundations

It is another fundamental of value theory that no final absolute proof can be advanced to support the values of one person or culture over those of another. The logical defense of any set of values will be found to rest ultimately on some axiomatic concept for which there is no proof and which must be accepted on the basis of faith or as being self-evident. In this respect, values are like the laws of physics, mathematics, and geometry—they rest on basic axioms that are accepted without proof. Even values of the intuitive, irrational variety may be shown at least to imply the acceptance of certain starting premises and innate biases. It follows directly that the basic postulates or starting axioms around which any system of values is built are critical in determining the total structure of the system.

In connection with the relativity and goal-dependency of values, it should be remembered that nothing has meaning in and of itself. A thing or concept is perceived and gets meaning and value only in terms of a background, a surround, something beyond or different from itself. Jumping ahead from these points, we find that it is assumptions concerning the meaning and goal of life as a whole, that have to be taken without proof, that in the last analysis stand behind most of those conflicting ideological and social values that now obstruct progress in crisis areas. It is of no surprise from an engineering standpoint that basic postulates accepted without supporting evidence and located in a potent key position in a dominant control system should turn out to be the strategic flaw in the global chain of control.

The diagnostic search for the root causes behind today's crisis problems narrows progressively from the biospheric effects of human intervention generally, to the underlying human value factors, to those values of the cognitive, acquired category, to focus finally on the starting axioms and premises, explicit or implied, on which these cognitive values rest. It is these latter axioms, self-evident truths, articles of faith, etc., concerning ultimate values, that structure the social priorities which shape the written and unwritten codes that govern the human actions and decisions, which in turn determine the future on planet Earth. Any consensus for change in the starting propositions in this chain of control—as in a new Bill of Rights, a new set of Commandments, a revised Manifesto, an amended Constitution—promptly modifies the entire value structure.

An active approach to value problems can thus be narrowed from concerns with values in general to a much more pointed and strategic effort directed at the fundamental starting concepts and beliefs concerning the ultimate goals and values, expressed or implied, around which cognitive value systems are organized. When these ultimate concepts and beliefs are in error, then social values will be correspondingly out of line, and all the associated human endeavor will be misdirected accordingly. Many of the basic assumptions and beliefs that have shaped human value systems in the past were formulated far back in history, promulgated in an intellectual climate in which it was supposed that the world was flat, the sun circled the earth, the seat of the mind was in the liver, and there was unlimited room for human expansion. Creative growth and correction in these original concepts have been suppressed by the tendency of most institutionalized belief systems to demand that their fundamental faiths and beliefs not be questioned. Application of the self-corrective principles and approach of science in this area would help to assure that man's fundamental guidelines do not go awry.

None of this is to suggest that authority for society's values be turned over to science or to scientists as individuals. The suggestion, rather, is for a fusion of science with ethics and religion that would open our value-belief systems to free scientific inquiry and empirical examination in general, in order that the same kinds of rigorous principles demanded for reaching belief in science might be applied also in the realm of values. It would mean, in essence, that in dealing with value questions the inner mental processes of the brain should regularly be forced to check and double-check with outside reality. This is the fundamental law underlying the scientific method—a point that seems simple but is sometimes overlooked in statements on the essence of science (51). All the vast superstructure of technology and research, the rigorous quantification, order, and institutionalization of science visible to the layman, are merely elaborations that stem from this basic operating policy. It stipulates that in reaching conclusions, the workings of the brain, intuitive, rational, emotive, etc., are not to be trusted to proceed on their own but must be checked regularly to assure conformance with outside reality. The mind can reach belief by various routes; the route of science is distinguished by its rigorous demand that any beliefs must double-check with the empirical evidence and this-world reality. Despite frequent shifts and corrective revisions in scientific theory, the overall track record of the scientific method as a means of cosmic understanding remains unmatched. It is not necessary to presume to find final, absolute answers—only improved ones.

Concepts of Consciousness Are Critical

Doctrine regarding ultimate values is closely tied to beliefs about the properties of the human psyche or conscious mind and its relation to physical reality. Value codes based on reincarnation, afterlife or otherworld existence, cosmic and/or divine intellect, immortality, and the like, all imply preconceived answers. So long as the nature of mind and the mind-body relation remained shrouded in mystery, the spectrum of possibilities was almost unlimited; the whole problem of human values floated in a wide open sea of uncertainty where science could hardly get started, and value-belief systems of necessity had to be built on conjecture, intuition and revelation.

Advances in the mind-brain sciences of the last few decades have very substantially narrowed the latitudes for speculation. In particular, the accumulating evidence in neuroscience leads overwhelming today to the conviction that conscious mental awareness is a property of, and inseparably tied to, the living brain. This is something that modern science points to as a salient reality of our world that we now must face. Like the reality of evolution and the earth's rotation, this new knowledge of the neurosciences must be taken into account and our values and ultimate goals shaped accordingly. A concept of mind and matter emerges that supports a unifying this-world view of man in nature as a framework for value guidelines (74, 77). Perhaps more than any other single development, the advances of the last half-century in our understanding of the neural mechanisms of mind and conscious awareness clear the way for a rational approach in the realm of values. This is not to say that the problems of mind-brain relations are now solved, far from it, but only that by process of elimination the range of realistic answers and their implications become much more rigorously defined.

Regardless of whether the answers of today in this or other areas prove final, the all-important ground rule for a scientific approach is the demand for concordance with verified evidence, whatever its status. This excludes values based on supernatural beliefs, any form of mystical insight, revelation, or unproven hypotheses about economics and class power struggles, however appealing these may appear to be. These limitations of the scientific way are at the same time its strength. Value-belief guidelines arrived at on these terms can be more reliably counted on to be protected from flaws of the kind that have undermined belief in the past.

Modern civilized society, with its greatly magnified global impact, unlike tribes in the jungle or even nations in centuries past, is under pressure to choose its value guidelines with a new kind of care and wisdom, around something of a higher order than man's natural reactions toward self-preservation or wish fulfillment. A new transcendent frame of reference is needed that cuts across all cultures, faiths, and national interests, for the welfare of mankind and the biosphere as a whole. Although this might eventually evolve spontaneously with time, as global conditions worsen, an active focused attack with a strong assist from science could do much to speed the process. If science today would retain public confidence and support in a situation where value problems have become primary, it might be advised to reverse openly its century-old rejection of human values as something outside its province. Few things could promise a more profound and widespread influence on the future than to bring together science and religion and other value disciplines in a crash program to better understand the origins and structure of human value systems, as well as what the world view of today's science could contribute to these value disciplines and to man's current search for a new improved ethic and sense of higher meaning.

On Implementation and the "Grand Design"

To determine in practical terms what consequent changes might be expected in mankind's sense of value becomes a separate undertaking of some complexity, calling for the combined interpretational insight of leaders from many disciplines, and going far beyond the limits of our present argument. However, a few brief words of preliminary nature may be helpful in response to the immediate doubts and questions that initially arise concerning the quality and kinds of values with which society might find itself afflicted on the above terms. How will values founded in the cosmic view and realism of science stack up in comparison with those based on supernatural and otherworldy views? Initial apprehension in regard particularly to spiritual richness and appeal seems to largely disappear upon exploration, provided (and this must be underlined) the fallacies of scientific reductionism are avoided as further explained in the chapter that follows.

Consider, for example, as a tentative starting maxim for determining right and wrong, to be accepted axiomatically without proof, something like the following: "The grand design of nature perceived broadly in four dimensions to include the forces that move the universe and created man, with special focus on evolution in our own biosphere, is something intrinsically good that it is right to preserve and enhance, and wrong to destroy or degrade." With this start, defined strictly in terms that are scientifically sound, an extensive and coherent value-belief system can be constructed. Other axioms and propositions may be added as long as they are consistent. The kind of moral code that logically emerges will contain much in common with alternative systems based on otherworldly beliefs, intuition, Christian, Buddhist, Communist, and other doctrine. Beyond the basic commonality, however, significant fringe differences become evident that are critical for current world problems. Ancient taboos, mythical beliefs, and a variety of local traditions, barbarisms, and sacred cows disappear as value determinants. A new perspective and new emphasis are implied concerning issues relating to population control, pollution, plunder of the ecosystem, species' rights, and related questions.

Individual freedom of choice, flexibility, and diversity may be inferred in regard to personal values as long as these subset values are not in overt conflict. Man continues to occupy a top position of prime importance, and priorities for most of the higher qualities of human civilization would be preserved. Man would, however, in any system based in scientific realism, probably lose some of his former unique, absolute status as the one measure of all things. Human society would no longer be justified in destroying or downgrading the rest of creation for its own homocentric aims. A significant revision of value standards results also from the use of a broad evolutionary framework. Value-belief systems of the past have been rather strictly human-oriented, on the one hand, and divine-myth-oriented, on the other. Today's conditions call for long-term, biospheric perspectives in which this world is conceived to be more than merely a way station to something better beyond.

It may be noted that the "grand design" of the sample axiom includes, by definition, the trends of evolution. The upward thrust of evolution as part of the design becomes something to preserve and revere. This would imply a commitment to progress and improvement—not in the municipal chamber of commerce sense, but in terms of the evolutionary trend toward greater complexity, diversity, and improvement in the quality and dimensions of life and the life experience. A sense of purpose and meaning is thus provided for the life of the individual and for society as a whole.

It is important to emphasize that a starting postulate of the sort illustrated, though based in science, is not an irreverent one. Ultimate respect and reverence for the cosmic forces of creation that control the universe and created man (including the conscious self-creative apex of these forces in the upper reaches of the human mind) are retained in full; only the definition and conception are modified to conform with modern evidence. Instead of relating to a single, omnipotent, personalized control force, man would relate to a vast complex of moving forces, hierarchically interlocked from the subatomic through the cellular, organismic, cognitive, social, and even galactic levels, in a great pluralistic system of cosmic forces in which the higher transcend the lower and all are differentiated from, and united in, a common foundation. Much of the great humanistic teaching of the past would be little changed in its basic impact. The "grand design of nature" as seen through the expanding eyes of modern science would appear already, in its presently understood form, to contain as much to sustain the highest in man's religious and spiritual needs, as have some of the comparatively simple dualistic mythologies. Instead of a rigid, closed scheme, we are led to one that continues to unfold and enlarge indefinitely as science and understanding advance.

The practical consequences effected by a value shift of this kind can be seen to stretch out endlessly. Prevention of environmental pollution or ravishment of the ecosystem, for example, becomes more than a mere expedient for human benefit. The ultimate meaning and purpose of all life are at stake, and a corresponding conviction, conscience, and dedication to what is sacred come to reinforce the effort. Comparable changes are realized in respect to species' rights, optimization of human numbers, nuclear escalation, and the like. Present trends to the contrary, humanity needs to see itself in terms of something greater and more important than itself to give meaning and purpose to human existence. The social system, commune, or all of humanity in general is not enough. With prior forms of dualist mythology now widely rejected by the informed majority, something like the grand design of the sample maxim is needed.

2

Mind, Brain, and Humanist Values

> . . . and in the mind of man:
> A motion and a spirit, that impels
> All thinking things, all objects of all thought,
> And rolls through all things.
> —Wordsworth 1798

Accusations from Antiscience

When I am asked as a scientist to explore some of the humanist implications of our brain-behavior sciences, I often find myself feeling a little like the accused who has been asked to mount the stand in self-defense. As they say back in grade two these days, for every action there is an equal and opposing reaction; and the recent sharp boom in science has not come without a corresponding rise in the voices of antiscience. Some of the going complaints in this regard are no doubt familiar: It is not only that science is going to blow us all off the globe, or crowd us off with its programs for death control, but that even the good things resulting from science—the sum total of all the better-things-for-better-living—have failed, we are told, to add substantially to a genuine satisfaction in living. And when it comes to the more profound humanist concerns, the reasons for living and the meaning and the value of it all, science seems only to take away and destroy, they say, and then refuses on principle to answer for its actions or even to be concerned with matters of values.

To some, even the objective explanatory progress that science is supposed to be making toward truth and the great central mystery of the universe begins to look like merely a handy system of humanoid guesses and correlational probabilities with no real verification possible. Others liken our explanatory progress to the penetration of a great maze that gets ever bleaker, the innermost chamber of which, should it ever be reached, being likely to hold almost nothing or perhaps just the self-reflections of the scientists' own thought processes. And then, about as fast as our comprehension and control of nature goes up, antiscience sees man's rating in the grand design going down.

Before continuing, I had better explain that the reference to values above and in the title was not accidental, though I well realize that any mixing of values and science tends to serve as a red flag in many quarters. Some of us may already be wondering: "Since when do scientists presume to carry a license for discussions of values?" Value judgments, we have all heard, lie outside the realm of science. Value matters are for popes and prophets, for philosophers and perhaps Boy Scout and YMCA leaders, but not for science or scientists. As a student of brain and behavior, I have never been quite able to accept this. It seems the same as saying that value judgments lie outside the realm of knowledge and understanding. It is like saying that the best method we know of applying the human brain to problems of understanding must be discarded when it comes to problems of values. It is almost the same as saying that economics is riding under false colors in the National Science Foundation and ought to be exposed and expelled. And it is like saying that science is able to deal only with those phenomena and products of evolution that appeared prior to the emergence of higher brains, with their wants, needs, goal-directed properties and, of course, the corresponding value systems that these impose.

Values have natural and logical origins. They are interdependent and interrelated in logical, hierarchical systems. These systems, and the perturbations thereof, ought to be subject to study and analysis and perhaps prediction, and even some experimentation on a model basis these days, with computer assistance. I have always wondered whether rather little harm and perhaps much good in the long run might not come from opening to the free winds of scientific skepticism and inquiry even the most revered of our traditional and cultural values.

Humanist Impacts of the Mind-Brain Sciences

We can now turn to our main question: What have been the major *humanist* impacts, up to the mid-sixties, of the recent developments in the sciences that deal with mind and brain? At first glance the record achieved by the brain-behavior sciences during the preceding half-century must seem to the humanist to read less like a list of contributions and advancements than like a list of major criminal offenses. The accusations that antiscience can raise in this area are not exactly trivial. For example, before science, man had reason to believe that he possessed a mind that was potent and full of something called "consciousness." Our modern experimental objective psychology and the neurosciences in general would divest the human brain of this fantasy and, in doing so, would dispense not only with the conscious mind but with most of the other spiritual components of human nature, including the immortal soul. Before science, man used to think that he was a spiritually free agent, possessing free will. Science tells us free will is just an illusion and gives us, instead, causal determinism. Where there used to be purpose and meaning in human behavior, science now shows us a complex biophysical machine with positive and negative feedback, composed entirely of material elements, all obeying the inexorable and universal laws of physics and chemistry. Thanks to Freud, with some assistance from astrophysics, science can be accused further of having deprived the thinking man of a Father in heaven, along with heaven itself. Freud's devastating indictment is said by many to have reduced much of man's formalized religion to little more than manifestations of neurosis.

Man's inner self and his heritage have not fared much better. Thanks to Darwin, and to Freud again, man now enters this life, not "trailing clouds of glory," as the poet once had it, but trailing instead clouds of jungle-ism and bestiality with a predisposition to Oedipal and other complexes. The veneer of civilization is seen to be superficial, and when it rubs thin or cracks, the basic animal within quickly shows. In the face of these and related onslaughts of science on the worth and meaning of human nature and existence, one can understand why humanist thinkers look for other roads to truth. For the scientist himself, the current dim picture puts a rather severe test to his credo that it is better to know and live by the truth, however disappointing, than by false premises and illusory values.

I find that my own conceptual working model of the brain leads to inferences that are in direct disagreement with many of the foregoing; I must take issue especially with the whole materialist-reductionist conception of human nature and mind that seems to emerge from the currently prevailing objective analytic approach in the brain-behavior sciences. When we are led to favor the implications of modern materialism in opposition to older idealistic values in these and related matters, I suspect that we have been taken, that science has sold society and itself a somewhat questionable bill of goods. There is not space here to present the whole story behind these remarks, and so I will try to concentrate selectively on what would seem to be the centermost issues, hoping that if the central foundation of the materialist view can be undermined, the resultant crumbling in the upper structures will become evident.

Nature of Consciousness: Central Issue

Most of the disagreements that I have referred to revolve around, or hinge either directly or indirectly upon, a central point of controversy that emerges from the following question: Is it possible, in theory or in principle, to construct a complete, objective explanatory model of brain function without including consciousness in the causal sequence?

If the prevailing view in neuroscience is correct, that consciousness and mental forces in general can be ignored in our objective explanatory model, then we come out with materialism and all its implications. On the contrary, if it turns out that conscious mental forces do in fact govern and direct the nerve impulse traffic and other biochemical and biophysical events in the brain and, hence, do have to be included as important features in the objective chain of control, then we come out at the opposite pole, with mentalism, and with quite a different and more idealistic set of values all down the line. We deal here, of course, with the old mind-body dichotomy, the age-old problem of mind versus matter, the issue of the spiritual versus the material, on which books and books have been written and philosophies have foundered ever since man started to think about his inner world and to question its relation to the outer "real" world.

Let us begin by stating the case against consciousness and mind as raised by today's objective experimental psychology, psychobiology, neurophysiology, and the related disciplines. The best way to deal with consciousness or introspective, subjective experience in any form, these disciplines tell us, is to ignore it. Inner feelings and thoughts cannot be measured or weighed; they cannot be centrifuged or photographed, chromatographed, spectrographed, or otherwise recorded or dealt with objectively by any scientific methodology. As some kind of introspective, private, inner something, accessible only to the one experiencing individual, they simply must be excluded by policy from any scientific model or scientific explanation.

Furthermore, the neuroscientist of today feels he has a pretty fair idea about the kinds of things that excite and fire the nerve cells of the brain. Cell membrane changes, ion flow, chemical transmitters, pre- and post-synaptic potentials, sodium pump effects and the like, may be on his list of acceptable causal influences—but not consciousness. Consciousness, in the objective approach, is clearly made a second-rate citizen in the causal picture. It is relegated to the inferior status of an inconsequential by-product, an epiphenomenon, or most commonly, just an inner aspect of the one material brain process. Scientists can see the brain as a complex, electrochemical communications network, full of nerve impulse traffic and other causally directed chemical and physical phenomena, with all elements moved by respectable scientific laws of physics, chemistry, physiology, and the like; but few are ready to tolerate an interjection into this causal machinery of any mental or conscious forces.

This is the general stance of modern behavioral science out of which comes today's prevailing objective, mechanistic, materialistic, behavioristic, fatalistic, reductionistic view of the nature of mind

and psyche. This kind of thinking is not confined to our labora-
tories and classrooms, of course. It leaks and spreads, and though
never officially imposed on the societies of the Western world, we
nevertheless see, everywhere we turn, the pervasive influence of
creeping materialism.

Once we have materialism squared off against mentalism in this
way, I think we must all agree that neither is going to win the
match on the basis of direct, factual evidence. The facts simply do
not go far enough to provide the answer, or even to come close.
Those centermost processes of the brain with which consciousness
is presumably associated are simply not understood. They are so
far beyond our comprehension at present that no one I know of
has been able even to imagine their nature. We are speaking here
of the brain code, the physiological language of the cerebral
hemispheres. There is good reason to believe that this language
is built of nerve impulses and related excitatory effects in nerve
cells and fibers and perhaps also in those glia cells that are said to
outnumber the nerve cells in the brain by about ten to one. And
we would probably be safe in the further noncommittal statement
that the brain code is built of spatiotemporal patterns of excitation.
But when it comes to even imagining the critical variables in these
patterns that correlate with the variables that we know in inner,
conscious experience, we are still hopelessly lost.

Furthermore, the central unknowns directly associated with
consciousness seem to be rather well cushioned on both the input
and output sides of the brain by further zones of physiological
unknowns. Our explanatory picture for brain function is reason-
ably satisfactory for the sensory input pathways and the distal
portion of the motor outflow. But that great in-between realm,
starting at the stage where the incoming excitatory messages first
reach the cortical surface of the brain, still today is very aptly
referred to as the "mysterious black box."

To conclude that conscious, mental or psychic, forces have no
place in filling this gap in our explanatory picture is at least to go
well beyond the facts into the realm of intuition and speculation.
The doctrine of materialism in behavioral science, which tends to
be identified with a rigorous scientific approach, is thus seen to
rest, in fact, on an insupportable mental inference that goes far

beyond the objective evidence and hence is founded on the cardinal sin of science. One can still find here and there in the literature a modicum of some final, perhaps "last rite," respect paid to the psyche. For example, there is the acceptance by Charles Sherrington of the possible coexistence of two separate phenomenal realms in the brain, and there is the stand of Carl Rogers that man's inner experience must be recognized as well as the brain mechanism of objective psychology. In the existence of two such very different realms, Rogers sees a lasting paradox with which we all must learn to live. But even the dualists are quite prepared to go along these days with the conviction held by most brain researchers—up to some 99.9 percent of us, I suppose—that conscious mental forces can be safely ignored, insofar as the objective, scientific study of the brain is concerned.

An Alternative Mentalist Position

In the pages that follow, I am going to align myself with the 0.1 percent or so mentalist minority in a stand that admittedly also goes well beyond the facts. It is a position, however, that seems to me equally strong and somewhat more appealing than those we have just outlined. In my own hypothetical brain model, conscious awareness does get representation as a very real causal agent and rates an important place in the causal sequence and chain of control in brain events, in which it appears as an active, operational force. Any model or description that leaves out conscious forces, according to this view, is bound to be sadly incomplete and unsatisfactory. The conscious mind in this scheme, far from being put aside and dispensed with as an inconsequential "by-product," "epiphenomenon," or "inner aspect," as is the customary treatment these days, gets located instead front and center, directly in the midst of the causal interplay of cerebral mechanisms. Mind and consciousness are put in the driver's seat, as it were; they give the orders, and they push and haul around the physiological and the physical and chemical processes as much as, or more than the latter processes direct them. This scheme is one that puts mind back over matter, in a sense, not under or outside or beside it. It is a scheme that idealizes ideas and ideals over physical and chemical interactions, nerve impulse traffic, and DNA. It is the brain model in which conscious mental psychic forces are recognized to be the crowning achievement of some five hundred million years or more of evolution.

Now, what is the argument in favor of mentalism, the argument that holds that ideas and other mental entities push around the physiological and biochemical events in the brain? The argument is simple and goes as follows: First, it contends that mind and consciousness are dynamic, emergent (pattern or configurational) properties of the living brain in action. There are usually plenty of "takers" on this first point, including even some of the tough-minded brain researchers as, for example, the outstanding neuroanatomist, C. J. Herrick. Second, the argument goes a critical step further and insists that these emergent properties in the brain have causal potency—just as they do elsewhere in the universe. And there we have the simple answer to the age-old enigma of consciousness. Who was it who said that nothing is so simple as yesterday's solution, nothing so complicated as tomorrow's problem?

But let us spell out this answer a little further, since this whole subject has at times been a bit confusing and complicated. To put it very simply, it comes down to the issue of who pushes whom around in the population of causal forces that occupy the cranium. It is a matter, in other words, of straightening out the peck-order hierarchy among intracranial control agents. There exists within the cranium a whole world of diverse causal forces, as in no other cubic half-foot of universe that we know. At the lowermost levels in this system, we have local aggregates of some sixty or more types of subnuclear particles interacting with great energy, all within the neutrons and protons of their respective atomic nuclei. These entities, of course, do not have very much to say about what goes on in the affairs of the brain. We can pretty well forget them, because they are all firmly trapped and kept in line by their atomic overseers. The atomic nuclei and associated electrons are also, of course, firmly controlled in turn. The various atomic and subatomic elements are "molecule-bound"—that is, they are hauled and pushed around by the larger spatial and configurational forces of their encompassing molecules.

Similarly, the molecules of the brain are themselves pretty well bound up and ordered around by their respective cells and tissues. Along with all of their internal atomic and subnuclear parts and their neighboring molecular partners, the brain molecules are obliged to submit to a course of activity in time and space that is very largely determined, for the lifetime of any given cell, by the overall dynamic and spatial properties of the whole cell as an entity. Even the brain cells, however, with their long fibers and impulse-conducting properties, do not have very much to say about when they are going to fire their messages, for example, or in what time pattern they will fire them. The firing orders for the day come from a higher command.

In other words, the flow and the timing of impulse traffic through any brain cell, or even a nucleus of cells in the brain, are governed largely by the overall encompassing properties of the whole cerebral circuit system, within which the given cells and fibers are incorporated, and also by the relationship of this circuit system to other circuit systems. Further, the dynamic properties of the cerebral system as a whole, and the way in which these properties direct and govern the flow of impulse traffic throughout the system—that is, the general circuit properties of the whole brain—may undergo radical and widespread changes from one moment to the next with just the flick of a cerebral facilitatory "set." This "set" is a shifting pattern of central excitation that will open or prime one group of circuit pathways with its own special pattern properties, while at the same time closing, repressing, or inhibiting endless other circuit potentialities that might otherwise be open and available for impulse traffic. These changes of "set" are responsible, for example, for such things as a shift of attention, a turn of thought, a change of feeling, or a new insight. To make a long story short, if one keeps climbing upward in the chain of command within the brain, one finds at the very top those overall organizational forces and dynamic properties of the large patterns of cerebral excitation that are correlated with mental states or psychic activity. And this brings us close to the main issue.

We can take this argument a step further by looking at an illustrative example of one of these mental entities. For simplicity, let us consider an elemental sensation. Instead of philosophy's old favorite, the color red (the philosophic and geographic locus of which seems sometimes to be in some doubt), let us use another example, pain. To be more specific, let us say we are talking about pain in the fingers and thumb of the left hand, and let us pin it down further to pain in the left hand of an arm that was amputated above the elbow some months previously. You will recall that the suffering caused by pain localized mentally in a phantom limb is no easier to bear than that in a limb that is still there. It will be easier, however, by using this example, for us to infer where our conscious awareness really resides.

In regard to the pain in a phantom limb, my contention is that any groans it may elicit from our patient and any other response measures or behavioral outputs that may be taken to be the result of the pain sensation are indeed caused not by the biophysics, chemistry, or physiology of the cerebral nerve impulses as such, but by the pain quality, the pain property, per se. This brings us, then, to the real crux of the argument. Nerve excitations are just as common to pleasure, of course, as to pain, and the same is true of any other sensation. What is critical is that unique patterning of cerebral excitation that produces pain instead of something else. It is the overall functional property of this pain pattern as a pattern that is critical in the causal sequence of brain affairs. This pattern has a dynamic entity, the qualitative effect of which must be conceived functionally and operationally and in terms of its impact on a living, unanesthetized cerebral system. It is this overall pattern effect in brain dynamics that is the pain quality of inner experience. To try to explain the pain pattern or any other mental qualities only in terms of the spatiotemporal arrangement of nerve impulses, without reference to the mental properties and the mental qualities themselves, would be as formidable as trying to describe any of the endless variety of complex molecular reactions known to biochemistry wholly in terms of the properties of electrons, protons, and neutrons and their subnuclear particles, plus (and this, of course, is critical) their spatiotemporal relationships. By including the spatiotemporal relations, such a description becomes feasible in theory, probably, but fantastically impractical. Moreover, by the time science arrives at a point where it can describe the critical details of the impulse pattern of a mental experience in the functional terms and setting required, it will be describing, in effect, the conscious force or property itself. When we reach such a point, the conscious force will be recognized as such, and we will be calling it just that—or at least that is the hypothesis I am putting forward.

Mind Over Matter

Many readers will note my reliance throughout this discussion on the emergent concepts of Lloyd Morgan and others and the corresponding configurational and field concepts of Gestalt psychology. The Gestalt school of psychology and philosophy went wrong only when they moved into the brain and tried to transfer their pattern properties directly from the outside world and sensory surfaces into the cerebral cortex. The central, emergent conscious force within the brain, as visualized here, is not a simple surrounding envelope, or volume property, or any other kind of "isomorph," as the Gestalt school tried to make it. It is rather a functional pattern that has to be worked out in entirely new terms, that is, in terms of the functional circuitry of the brain, in terms of the still unknown brain code.

Above simple pain and other sensations in brain dynamics, we find, of course, the more complex but equally potent forces of perception, emotion, reason, belief, insight, judgment, cognition, and all the rest. In the onward flow of conscious brain states, one state calling up the next, these are the kinds of dynamic entities that call the plays. It is exactly these encompassing mental forces that direct and govern the inner impulse traffic, including its electrochemical and biophysical aspects. When trying to visualize mental properties as they have been described, it is important to keep in mind the fact that all of the simpler, more primitive, electric, atomic, molecular, cellular, and physiological forces remain present, of course, and they all continue to operate. None has been canceled, but these lower level forces and properties have been superseded, encompassed, as it were, by those forces of successively higher organizational entities. We must remember in particular that, for the transmission of nerve impulses, all of the usual electrical, chemical, and physiological laws still apply at the level of the cell, the fiber, and the synaptic junction. We must remember further that proper function in the uppermost level always depends on normal operation at subsidiary levels.

Near the apex of this command system in the brain—to return to more humanistic concerns—we find ideas. Man, unlike lower animals, has ideas and ideals. In the brain model proposed here, the causal potency of an idea, or an ideal, becomes just as real as that of a molecule, a cell, or a nerve impulse. Ideas cause ideas and help evolve new ideas. They interact with each other and with other mental forces in the same brain, in neighboring brains, and thanks to global communication, in far distant, foreign brains. And they also interact with the external surroundings to produce in toto a burstwise advance in evolution that is far beyond anything to hit the evolutionary scene yet, including the emergence of the living cell.

In the proposed scheme, the interplay of psychic and mental forces, though accessible—like the interior of the earth—only indirectly at this date, becomes, in principle, a proper phenomenon for scientific investigation. Aside from problems of complexity and adequate technology, there would seem to be no great obstacle in principle to the eventual objective, scientific treatment of mental phenomena. One may see statements in the literature these days discouraging the hope that the mind is capable of explaining itself in terms of its own ideas; the argument used is that no machine, living or otherwise, can logically embody within itself a complete description of itself. When you read such statements, however, always underline that word "complete" and then consider the extent of the explanatory possibilities that still remain even though they fall somewhat short of complete. Underline also that word "itself" and remember that this kind of logic does not prevent a man's mind from acquiring a complete description of his neighbor's mind or from passing on this description to other neighbors, excepting only the one he has described.

Split Brain, Split Mind

For an outside, second brain to directly experience the subjective qualities in an observed brain, it would seem to be necessary for the second brain in the observer to be coupled in parallel to the observed brain and wired directly into the specialized cerebral circuitry involved. This does not seem very feasible under ordinary conditions for the near future. However, we do seem to be approaching exactly this situation experimentally in recent studies in which the brains of cats and monkeys have been bisected down the midplane into right and left halves. In the surgical process, a few cross-connections may be left, coupling selected cerebral centers between "mind-right" and "mind-left." When the midline disconnection is complete, two separate mentalities are the result, which sense, perceive, learn, and remember independently. Each half seems to have its own realm of conscious awareness, and each is apparently as much out of contact with the inner mental images of the other as are two brains in separate skulls. But when a band of cross-connections is left intact, linking, for example, the right and left centers for vision or those for touch sensibility in the hands, the inner, mental, subjective experience of the one brain seems to become available to the other.

The same phenomenon can also be seen in studies of human patients who have had a similar surgical disconnection of the hemispheres for medical or therapeutic purposes and in whom cross-connections have been left intact between the lower brain centers involved in emotion and feeling. Whereas the cognitive, perceptual, mnemonic, and related experiences of mind-right in these people seem to be entirely out of touch with the correspond-ing experience of mind-left, each brain half seems to share the emotional experience of the other. For example, if an emotion is triggered through vision by the introduction of an unexpected pinup picture of a nude into a sequence of ordinary geometric pattern stimuli being projected into only the right lobe, it is quite apparent from the verbal readout through the other half of the brain (that is, the one not directly excited) that this second hemisphere also feels properly embarrassed—or whatever the case may be. The second hemisphere, however, has no idea why it has these inner feelings and is unable to describe their source.

Unifying World View

Looking back from this point, you will note that the earlier basic distinction or dichotomy between mentalism and materialism is resolved in this interpretation, and the former polar differences with respect to human values, when recast in the present scheme, become mainly errors of reductionism. This may be recognized as the old "nothing but" fallacy; that is, the tendency, in the present case, to reduce mind to nothing but brain mechanism, or thought to nothing but a flow of nerve impulses. For those acquainted with theories of mind, the new twist here is to be found in the attempt to make the emergent properties of inner experience conform to the inner brain code, rather than to the outside world or subjective impressions or sensory patterns, combined, of course, with the critical interjection of these mental qualities into the causal sequence. Note that we have not rejected the objective approach of science; it is an objective explanatory model that we are discussing. Our quarrel is not with the objective approach but with the long-accepted demand for exclusion of mental forces, psychic properties, and conscious qualities from the objective scientific explanation.

The present scheme would put mind back into the brain of objective science and in a position of top command. If correct, it would eliminate the old dualistic confusions, the dichotomies and the paradoxes, proposing instead a single unified system extending from subnuclear forces near the bottom up through ideas at the top. As a scientific theory of mind, it would provide a long sought unifying view on which to base our conception of human nature, the kind of view that humanists long have needed and lack of which has also been deplored recently in lead articles in *Science*. Moreover, this scheme suggests a possible answer not only for the relation between mind and brain but also for that between the outside world and its inner cerebral representation, another conundrum since the days of Plato. When used as a conceptual skeleton on which to build a body of philosophy, this theory favors a single this-world measuring stick for evaluating man and existence. As for the older materialist doctrine, one can say, in summary, that the denial or downgrading of conscious mental forces in objective experimental psychology during the past half-century may have been valuable as a tactical expedient for a developing science, but it is hardly a doctrine on which to build social philosophy or cultural values.

Mental Self-Control and Free Will

Another serious threat to cherished images of human nature is the scientific rejection of free will. Every advance in the science of behavior, whether it comes from implanted electrodes, psychomimetic drugs, the psychiatrist's couch, brain surgery, imprinting, or Skinner boxes, seems only to reinforce the old suspicion that free will is merely an illusion. The more we learn about the brain and behavior, the more deterministic, lawful, and causal it appears. Attempts to restore free will to the human brain by recourse to various forms of indeterminacy,—physical, logical, emergent, or others—have failed, so far as I can see, to do much more than perhaps introduce a bit of unpredictable caprice into our comportment that most of us would prefer to be without. Neither science nor philosophy seems able as yet to find in the brain any satisfying exceptions to the onward flow of causal determinism.

Before we become overly disturbed by all this, however, there are a few more points we should keep in mind. These add up to the conclusion that if we were given freedom of choice in this whole matter, we might well prefer not to have it; that is, we would probably prefer to leave determinism in control exactly as science postulates. It should be clear that the kind of determinism proposed is not that of the atomic, molecular, or cellular level, but rather the kind that prevails at the level of cerebral mentation, involving the interplay of ideas, reasoning processes, judgment, emotion, insight, and so forth.

The proposed brain model provides in large measure the mental forces and abilities to determine one's own actions. It provides a high degree of freedom from outside forces as well as mastery over the inner molecular and atomic forces of the body. In other words it provides plenty of free will as long as we think of free will as self-determination. A person does indeed determine with his own mind what he is going to do and often from among a large series of alternative possibilities.

This does not mean, however, that there are cerebral operations that occur without antecedent cause. Man is not free from the higher forces in his own decision-making machinery. In particular, our model does not free a person from the combined effects of his own thought, his own impulses, his own reasoning, feeling, beliefs, ideals, and hopes, nor does it free him from his inherited makeup or his lifetime memories. All these and more, including unconscious desires, exert their due causal influence upon any mental decision, and the combined resultant determines an inevitable but nevertheless self-determined, highly special, and highly personal outcome. Thus the question: Do we really want free will, in the indeterministic sense, if it means gaining freedom from our own mind, from our own self and inner being?

There may be worse fates, perhaps, than causal determinism. Maybe after all it is better to be an integral part of the causal flow of cosmic forces than to be out of contact with these—free-floating, as it were, with behavioral possibilities that have no antecedent cause, and hence no reason nor any reliability relative to future plans, predictions, or promises. If one were assigned the task of trying to design and build the perfect freewill model, consider the possibility that the aim might be not so much to free the machinery from causal contact as the opposite; that is, to try to incorporate into the model the potential value of universal or unlimited causal contact. In other words, contact with all related information in proper proportion—past, present, and future.

At any rate, it is clear that the human brain has come a long way in evolution in exactly this direction, when you consider the amount and the kind of causal factors that this multidimensional, intracranial vortex draws into itself, scans, and brings to bear in turning out one of its "free choice" decisions. Potentially included, through memory, are the events and wisdom of most of a human lifetime. Potentially included, also, with a visit to the library, is the accumulated knowledge of all recorded history. And we can add, thanks to reason and logic, much of the predictive value extractable from all these data as well as creative insights newly conceived. Maybe the total falls a bit short of universal causal contact; maybe it is not even up to the kind of thing evolution has going for it over on galaxy nine; and maybe, in spite of all, any decision that comes out is still predetermined. Nevertheless, it certainly represents a very long jump in the direction of freedom from the primeval slime mold, the Pleistocene sand dollar, or even the latest model orangutan.

It will be evident that our current view does not deny the animalistic in human nature—any more than it denies the molecular or atomistic. It does deny, however, that the higher human properties in the mind and nature of man are the same as, or are reducible to, the components from which they are fashioned. On the debit side, there is little in our proposed model for consciousness to bolster one's hopes either for extrasensory perception or for postmortem perception. Similarly, prepartum perception in the embryo would presumably be negligible until after the requisite cerebral machinery for conscious awareness begins to attain functional maturity in the later months of fetal life, and in subsequent postnatal development.

Heredity or Experience?

Finally, in connection with fetal development let me mention briefly, before closing one other area where science for decades appears to have been giving us some major misconceptions regarding human nature. This last concerns the old nature/nurture problem: the extent to which behavior traits are inherited versus the extent to which they are acquired and subject to shaping by experience and environment.

Through most of the first part of this century and up until about twenty years ago, the view prevailed that the brain gets its start in fetal life as an essentially equipotential network, a blank slate, as it were, which is then gradually channeled from early fetal movements onward by functional trial and error, conditioning, practice, learning, and experience. The objective, materialist movement in psychology, early influenced by the work and ideas of Pavlov in Russia, and pioneered in this country by Watson under the name "behaviorism," has been identified almost as much with the promotion of the conditioned response as it has with the demotion of consciousness.

The mind was thought to develop gradually out of a lifelong chain of successive conditioned-reflex associations, starting in the infant from a few elementary reactions, like love and hate, fear and anger. The whole idea of the genetic inheritance of behavior patterns came to be forcibly renounced. The term "instinct" became highly discredited in professional circles, its rejection almost equaling in fervor that of consciousness. The rejection of instinct was supported by the belief in those days that the embryonic growth of brain pathways was nonselective and diffuse, and the establishment of precise fiber connections was held to be unimportant anyway for orderly function. The brain connections, once laid down, were thought to be capable of radical wholesale rearrangement of function to compensate for surgery, injury, or misregeneration. In the scientific thinking of those times, the brain was endowed with an almost mystical, omnipotent plasticity and readaption capacity. In general, science seemed to be telling us through the 1920s, '30s, and early '40s that the human brain and human nature in general were extreme in their malleability. It seemed at that time a scientifically sound deduction that with an appropriate program of training and environmental conditioning it should be possible to shape human nature and hence society, within wide limits, into a desired mold.

Much of the basic scientific thinking and evidence behind this view has since suffered a series of severe upsets, leading to a current stand that is in many respects diametrically opposed to the earlier doctrines. Instead of a loose, universal plasticity in brain hookups, we now find a basic built-in wiring plan characteristic of the species that is highly resistant to functional readjustments. Instead of diffuse, nonselective growth of nerve connections in brain development, we now have evidence for a highly ordered growth and patterning of brain pathways and fiber hookups, all strictly regulated with extreme precision through genetic control

reductionism, and mechanistic determinism have been corrected along lines explained in the preceding chapters. Hereafter, wherever I support science or scientific truths in relation to value priorities, it is not the old materialist view that is referred to but rather the new holist-mentalist paradigm, which has an entirely "new look" and stands for something that is no longer in conflict with ethical, religious, or other humanist sensitivities. Recall also that the content and concerns of the political, social, and other behavioral sciences are to be included in principle as well as the more basic sciences. I want to include here also any empirical knowledge that is as reliable as that garnered by the scientific method, for example, verified historical facts.

A number of different but convergent lines of reasoning, one of which I outlined in chapter 1, can be seen in support of this argument. All logically lead to and support the same general conclusion, namely, that it is quite feasible and within the bounds of reasonable hope to build a viable value system or code of ethics solidly founded in the kinds of truths and beliefs acceptable to science and credible for all mankind—without resorting to differing dualist constructs that only some peoples accept and others find intolerable. The approach I will use here is something of a shortcut and goes as follows: the supreme ultimate authority, arbiter, reference, or determinant of what is ethically and morally good, right and true, that has been used most widely and most commonly recognized throughout history, has been the concept, in various forms, of man's creator, along with the cosmic forces that move and control the universe. Visualized and defined in different ways to pagans, Christians, Jews, Buddhists, Moslems, Hindus, American Indians, and so on, this essential broad concept, as an ultimate arbiter and reference, can be seen to run as a common thread through most of man's great faith-belief systems. There are, of course, exceptions such as humanism, hedonism, and communism, which use lower level reference frames anchored in the human species itself rather than in some "higher authority."

Despite the predominant agreement through history with regard to the broad general concept, feelings run high, as we all know, with respect to specific differences in the way the forces of creation and cosmic causation have been variously conceptualized and the resultant dogma, doctrines, and traditions that have grown up around each. Battles have been fought over these differences and historians remind us that religious wars are the bloodiest. The reason, in large part, is that the sense of higher meaning and the value of existence itself (and, therefore, of most everything else that counts) is directly at stake in reference to these ultimate determinants at the top of our value hierarchies.

In any case, the result, with which we are all too familiar, is that the world community has had to live with a series of diverse ultimate standards or measuring sticks for what is ethically right and wrong, with the supporters of each system devoutly claiming theirs to be *the truth* to be accepted on faith by the followers, in many cases as being absolute and beyond question. The value differences that derive from these various belief systems, as already emphasized, exert profound and pervasive effects on social and political decision-making that add up to what is probably one of the most, if not *the* most formidable of the divisive influences that now confront us and operate against world harmony and unity.

One does not find much hope that this situation will be remedied in the foreseeable future by the gradual dominance of one or another of the currently extant religious systems at the expense of the others; nor through the appearance and rise of any new alternative based on the teachings, visions and revelations of any single individual inspired leader. One suspects the world has become too sophisticated, too diverse and complex to be united by such a development. Nor does the Communist framework appear to provide the kind of answer that is needed here with its materialist philosophy and relatively narrow perspectives based on class power struggles of industrialized human society. Current world problems demand higher perspectives and a transcending frame of reference that includes the long-term welfare of the entire ecosystem.

A more promising alternative would seem to lie in the search for a new reference frame for global ethics along lines indicated in the preceding chapters where criteria for ultimate value are required to jibe with scientific reality. For Communist countries this would demand some updating in interpretations of science and what science stands for, including a philosophic shift from materialism to the new mentalism. For most other countries it would mainly mean replacing or reconciling various otherworldly guidelines with this-world existence. These changes need not affect the entire citizenry, only those leaders involved in forming global policy. Scientific standards for validity are chosen, not with the idea that scientific truth is absolute or infallible, but only on the belief that it represents the best and most reliable, credible and dependable *approach to truth* available.

Search for Meaning

The desirability of trying to obtain uniform agreement on some explicit frame of reference in these matters is evident in theory as well as practice. Along with other intrinsic origins of human values we recognize an inherent tendency of the human mind to seek to perceive meaning in itself, including higher meaning. This is coupled with an innate reasoning faculty and the combination leads to a rational cognitive upper structure in the value systems of man. It is within this cognitive, rational realm, which is distinctly human, that the major ideological, political, and religious differences are found. When it comes to ordering social priorities in this realm the kind of ultimate rational reference frame that is used becomes critical.

The kinds of changes that would be required to bring the diverse contempory creeds into accord with a reference frame based in the reality of science should mainly be corrective rather than exclusive, and might be accomplished to a considerable extent in each case by updating and reinterpreting aspects of doctrine that have remained fixed for many centuries while science has been advancing. In the eyes of modern science, to put it simply, man's creator becomes the vast interdependent and interwoven matrix of all evolving nature, a tremendously complex concept that includes all of the immutable and emergent forces of cosmic causation that control everything from high energy subnuclear particles on up to galaxies—not forgetting the causal properties that govern brains and behavior at the individual and at social levels. For all of these, science has gradually become our recognized prime authority, offering a cosmic scheme and view of life that renders most others simplistic by comparison.

In line with the foregoing an acceptance of scientific, this-world constraints for ethics would lead to the designation in very broadest terms of what is good, right, or to be valued morally, as that which is in harmony with, sustains, or enhances the orderly design of evolving nature *including its human apex.* The "highest good" becomes expressed in terms of fitting in and contributing to the grand design of the creative process, i.e., furthering the progressive overall improvement in the diversity, meaning, and quality of existence. Conversely, that which is discordant, degrades, or destroys creation's overall pattern (by which the quality, beauty, and meaning of existence have been maintained and improved over eons), becomes wrong, evil. The reference is not to the innumerable subsystems within subsystems of the natural order but to the overall "grand orderly design" in long-term perspective with special focus on evolution in our own biosphere and its crowning peak in the mind of man. It is an overall pattern of ever higher levels in the quality of existence, higher and higher dimensions of conscious experience, and, in the human realm, progressively higher levels of aesthetic and spiritual awareness.

Right and Wrong from This-World Criteria

Though it may be premature at this point and overly optimistic, it does not hurt to try to foresee, at least in general outline, some of the kinds of value changes that might logically emerge on the above terms. With recognition of the role of evolving nature in the creation of man, one can infer directly an increased respect and reverence for all nature and for what is sometimes referred to as the "infinite wisdom of nature." The quality, balance, and progressive differentiation of the ecosystem as a whole would acquire heightened priority. Things like the recycle philosophy, species' rights, conservation of natural resources, and control of population explosions become reinforced with a higher kind of dedication and commitment beyond that of mere human expedience.

Mankind, as part of evolving nature and as the apex and main "growth tip" of the evolutionary scheme, remains the prime concern, but stands to lose some of the prior uniqueness and "measure-of-all-things" status accorded in some previous systems. In cases of conflict particularly, where human ideas of what is best obviously run counter to the time-tested principles and proven plan of all creation as a whole, greater deference is given to the latter. A sense of higher meaning can be preserved through a meaningful relation to something deemed more important than the human species taken by itself.

It is to be expected that one may find at first, in the early stages, many apparent difficulties, contradictions, and differences of opinion. In very broad terms it would mean for many of us the relinquishment of various otherwordly or dualistic guidelines in favor of the principles and embodied wisdom of the evolving natural order that have functioned successfully in evolution for ages and succeeded in creating man. Difficulties and contradictions are to be expected in applying any system of ethics used to date. The Judeo-Christian ethic, for example, is full of moral contradictions. The aim is not to eliminate all controversy and differences of opinion but only to bring these within a given explicit domain set by an agreed-upon frame of reference adhered to because it is believed to be the most valid. From the standpoint of the individual it it is important to note that a shift as described to a reference frame based on this-world validity would neither eliminate nor greatly alter the bulk of day-to-day personal values throughout lower levels of the value hierarchy. The biologically based family and many social interactions and relationships retain the same meaning and value.

Finally, in view of the overwhelming complexity of the forces of creation and cosmic causation and the consequent difficulty of an adequate direct conception, plus the natural, innate tendency of the human brain to personify hidden causation and to look to personal leadership for security, it becomes reasonable at times, and often more convenient for thinking and communication to personify the concept of the ultimate reference. The proviso must be made, of course, that any such personification should be remembered for what it is and not taken so literally that it leads to false inferences regarding the nature and properties of what has been personified and thus again to value priorities that are out of tune with reality.

4

Messages from the Laboratory

One would like to know more where really our values come from. And so you can ask where do the values come from, and you can ask what should our values be, and if you have an answer to what our values *should* be, how do we get them to be our values. These are not questions of science, but they are the questions the answer to which will decide the further course of history more than anything else. I think the further course of history will not be decided by further discoveries in science, but by these questions about human values.

—Max Delbrück 1980
Engineering and Science interview

Increasingly in recent times we scientists get asked by federal agencies and others to explain what there is in our research that may be "relevant" to current social needs. While such requests for relevance are often disconcerting in respect to basic science, I do find it possible to pick out at least three facets of our research work where the kind of message we have been getting from the laboratory seems to differ notably from the conventional story current in the public media or in society at large. The first two of these relate to basic matters that only indirectly concern questions of moral priority through their implications regarding man's inner nature and the innate development of the value structure.

Our Inherent Individuality

The first of these messages from the laboratory is concerned with the functional plasticity of brain organization and hence of behavior and of human nature in general. Back when I first began to work in this area, neuroscience was thoroughly sold on the idea that brain function was almost unlimited in its plasticity. Among other things, the functional interchangeability of nerves in the practice of nerve surgery was taken for granted. Having its wires crossed by the neurosurgeon was considered to be no problem at all for the brain back in the 1930s.

For example, when a damaged nerve like that supplying the muscles of the face was replaced surgically by a nearby healthy and more expendable nerve—such as that for lifting the shoulder—the initial effect was the occurrence of associated movements in the face whenever the subject tried to lift his shoulder. However, the doctrine of the day said that the patient need merely go home and practice in front of a mirror and shortly the plastic brain centers would undergo re-education to restore normal facial expression, mediated now through the brain centers and nerves designed for shoulder movement.

Efforts were being made to restore function to legs paralyzed from spinal cord damage by substituting one of the main nerves of the arm, leaving intact its original connections to the brain centers. The arm nerve was dissected out full length, cut free at the distal ends and then tunneled under the skin, to be reconnected to the leg nerves to take over the function of the paralyzed limb. Only an early report appeared in the literature, not the final outcome of this effort—perhaps for reasons that are now understandable. However, exactly the same operation was later reported to be a functional success in experimental tests with rats during the 1930s. The motor, the sensory, even the spinal reflex functions of the animals' paralyzed hind legs were said to have been restored through the transplanted foreleg nerves still activated from brain centers of the forelimb—a result considered impossible today but which illustrates the prevailing neurological theory of the thirties— and the power of wishful thinking.

The brain and nervous system generally appeared in those days to be possessed of a wholesale behavioral plasticity or, as one authority put it, "a colossal adaptation capacity almost without limit." The followers of Pavlov in Russia and of John Watson in this country were speculating (justifiably it seemed) that it should be feasible with appropriate early training and conditioning techniques to shape human nature into most any desirable mold and thus to create a more ideal society.

This kind of thinking was reinforced by various other mutually sustaining views of the 1930s. In particular, the prevailing doctrine on nerve growth held that fiber outgrowth and the formation of nerve connections in the developing fetal brain was essentially diffuse and nonselective. The intricate patterning of brain pathways and hookups was ascribed to scheduled timing and mechanical factors and especially to later feedback effects of activity. In those days there seemed to be no way by which the complex nerve circuits for behavior could be grown in directly—that is, prefunctionally through inheritance, without shaping by experience. It was supposed that the selective channeling of brain connections was dependent on function and began way back in the earliest movements of the fetus in utero—continuing from then on through trial and error, conditioning, learning, and experience.

Our experimental findings during the 1940s brought, of course, a direct contradiction amounting to a 180-degree about-face on these matters. As we now know, transplanted nerves are not at all functionally interchangeable, the brain is not all that plastic, and the growth of nerve pathways and nerve connections in the brain is anything but diffuse and nonselective. The tremendously intricate patterning of the brain circuits for behavior was found to be mainly achieved by growth processes, prefunctionally, under genetic control, and carried out with great precision, involving an enormously complex system of preprogrammed cell-to-cell chemical affinities.

It is not just to recall old times that I go back through this earlier history. The point here is that the early erroneous views that became deeply entrenched all through the 1920s, 1930s, and well into the 1940s still to this day have not been completely shaken off in areas outside the biomedical sciences. The lingering aftereffects of the earlier plasticity doctrines may still be found in disciplines like psychiatry, anthropology, and sociology and also in society at large. In other words, the majority of us still have a tendency to underestimate the importance of the role played by genetic and other innate factors in the shaping of brain organization and behavior.*

This conclusion comes not only from the kind of work just mentioned. It has continued to be reinforced consistently from many other different angles. For example, in regard to cerebral dominance and handedness in man, the latest theory suggests a two-gene, four-allele model, with one gene determining which hemisphere of the developing brain will become language-dominant, and a second gene determining whether the preferred hand will be on the same side or opposite the language hemisphere. Counting the recessives and dominants, this gives nine different combinations of inherited gene types or genotypes for handedness and cerebral dominance in man, some of the more strongly left-handed types being much more resistant, of course, than others to reversal by training.

The left and right hemispheres of the brain are each found to have their own specialized forms of intellect. The left is highly verbal and mathematical, and performs with analytic, symbolic, computer-like, sequential logic. The right, by contrast, is spatial, mute, and performs with a synthetic, spatioperceptual, and mechanical kind of information processing not yet simulatable in computers. It is very impressive and compelling in neurological patients with left and right hemispheres surgically disconnected to see the same person (some claim there are two persons in the one) approach the same problem, work it, and reach a solution in consistently different ways with quite different strategies, depending upon whether the subject is using his left or right hemisphere.

It follows that the above nine genotype combinations, representing different balancing and loadings of the left–right mental factors, provide just in themselves quite a spectrum for inherent individuality in the structure of human intellect. Left-handers as a group have been shown to be different statistically from right-handers in their mental makeup—that is, in their IQ and other test/profiles. Similarly, males come out differently than females.

* Strong new confirmation of the powerful role of inheritance has come recently from current studies still in progress on identical twins reared apart. See preliminary accounts in *Science* (1980), 207:1323; *Smithsonian* (1980), 11:48; and *Science 80*, 1(10):55.

And females masculinized in utero or those lacking one X chromosome come out differently than normal females.

Many kinds of tests have shown that the right hemisphere is particularly talented and superior to the left in visuospatial abilities. This specialty of the so-called minor hemisphere is tied to a recessive sex-linked gene and is shown to exhibit a cross-correlation pattern of inheritance from parents to offspring that effectively rules out environment, experience, or any known theory of child development or nurturance.

When we add up all this—and much more—related evidence, we come out with a greatly heightened respect and appreciation for innate individuality. The degree and kind of inherent individuality each of us carries around in his brain—in its surface features, its internal fiber and cell organization, microstructure, chemistry—would probably make those differences seen in facial features or in fingerprint patterns look crude and pale by comparison.

The Neglected Minor Hemisphere

We turn now to a second message that emerges from the findings on hemispheric specialization. For many decades it was supposed that the human brain evolved with a strong one-sided mental dominance on the left associated with language and higher cognitive abilities dependent on and related to language, along with the abstract, mathematical, logical, and other reasoning powers all thought to be located in the left hemisphere. The supposedly less evolved, rather backward and illiterate right hemisphere, on the other hand, was inferred to be not only mute and unable to write, but also "word deaf," and "word blind." It was reported further to be largely apraxic and agnostic, i.e., lacking in the capacity to understand and execute complex volitional movements and wanting in higher cognitive faculties generally.

These impressions arose historically from a long series of clinical studies of the effects of asymmetric brain damage. Right hemisphere lesions were much less disruptive of higher functions. On the other hand, even small focal lesions located in the left hemisphere might consistently produce any of the above disabilities despite the preservation of an intact right hemisphere. A lesion on the left, for example in Wernicke's area, or disconnecting this left brain area from the primary auditory regions, abolished the capacity to comprehend spoken language. Similarly, lesions of the left angular gyrus, or lesions merely interrupting its input from the visual areas, proved sufficient, as a rule, to abolish the capacity to read.

However, when we came to test patients who had undergone surgical disconnection of the hemispheres it turned out, much to our surprise, that the right hemisphere under these conditions was by no means either word deaf or word blind, nor was it apraxic or agnostic. The disconnected "minor" hemisphere in these "split-brain" or commissurotomy subjects was quite able to comprehend, at a moderately high level, words spoken aloud by the examiner. These subjects also were able to read printed words flashed to the left visual field as shown by selective manual retrieval or by pointing to corresponding objects or pictures in a multiple choice array. These patients were also able with the right hemisphere to choose correct written or spoken words to match presented objects and to go correctly from spoken to printed words and vice versa. Correct tactual retrieval by the minor hemisphere was achieved even for objects not directly named but desribed by the examiner with complex spoken phrases.

In further studies on the surgically separated hemispheres, involving direct right-left comparisons, where most of the usual confounding variables cancel out, the so-called subordinate or minor hemisphere has proven to be indeed the cognitively superior cerebral member in a large series of nonverbal performance tasks. These are also, of course, nonmathematical and nonsequential, largely spatial—the kind of tasks where a single spatial image processed as a whole proves more effective than a detailed verbal or mathematical description. Examples include the reading of faces, copying of designs, fitting forms into molds, the discrimination and recall of nondescript tactual and visual shapes, spatial transpositions and transformations, conceptual grouping of different sized and shaped blocks into categories, judging whole circle size from a small arc and perceiving whole forms from a collection of parts. Earlier doubts regarding the presence of advanced mental function in the minor hemisphere are now largely dispelled.

Everything we have observed over many years of testing reinforces the conclusion that the disconnected, mute hemisphere has an inner conscious experience of much the same order as that of the speaking hemisphere though different, of course, in quality. Clearly the right hemipshere perceives, thinks, learns, and remembers, all at a very human level. It also reasons nonverbally, makes studied cognitive decisions, and carries out novel volitional actions and complex learned movements. Further it is shown to generate typical human emotional responses when confronted with affect-laden stimuli and social situations.

Looking back now we must ask: Why does the right hemisphere succeed after surgical disconnection—or also after the occasional complete removal of the partner hemisphere—when it fails functionally after even small lesions in the partner hemisphere? The answer seems to be that the one-sided lesions tend to disrupt function not only on the damaged side but also on the undamaged side as well, through remaining commissural cross-connections. We must think of the two hemispheres as normally functioning together as an integral unit. When a given function is disrupted by a relatively small but critical damage, particularly on the side of most specialized control, the function of both hemispheres is therefore involved. Thus the lateral or asymmetric lesions on which the earlier doctrine had been based prove to be deceiving in respect to what the undamaged hemisphere can do.

In any case we have come around to accept today a substantially revised and upgraded picture of the right hemisphere and its functional capacities. The classic neurologic doctrine of one-sided dominance, with a major and a minor hemisphere, is replaced by the idea of a bilateral complementary specialization.

What I am getting to is that these developments regarding the right hemisphere seem to tell us, among other things, that our educational system and modern society generally (with its very heavy emphasis on communication and on early training in the three R's) discriminates against one whole half of the brain. I refer, of course, to the nonverbal, nonmathematical, minor hemisphere, which we find has its own perceptual, mechanical, and spatial mode of apprehension and reasoning. In our present school system, the minor hemisphere of the brain gets only the barest minimum of formal training, essentially nothing compared to the things that we do to train the left, or major, hemisphere. (As a curious aside here, statistics indicate that athletic abilities correlate with enhancement of the visio-spatial mental ability. It follows as an interesting conjecture that advancement in our understanding of the cerebral substrates of intellect could make for a slight comeback in the old prestigious image of the "strong silent man" of pioneer times—an image that is much submerged, of course, in our present-day verbal society.)

Behaviorism in Question

A third and final message for social change that we get from the world of the laboratory is a complex one and cannot be summarized simply. One of the more important things to come out of our brain research in recent years—from my own standpoint, at least— is a modified concept of the nature of the conscious mind and its relation to brain mechanism. The new interpretation, or reformulation, involves a direct break with long-established materialistic and behavioristic thinking that has dominated neuroscience for many decades. Instead of renouncing or ignoring consciousness, the new interpretation gives full recognition to inner conscious awareness as an important high-level directive force or property in the brain mechanism. The conscious mind no longer is set aside as a passive correlate, but becomes instead an essential part of the brain process endowed with causal potency. The phenomena of inner experience are conceived to be emergent properties of brain activity and become causal determinants in brain function.

On these new terms consciousness is given a use, a reason for being, and for having been evolved in a material world. Not only does the brain's neurophysiology determine the mental effects, as has generally been agreed, but now in addition the emergent mental operations are conceived in turn to control the component neurophysiology through their higher organizational properties and the universal principle of the power of the whole in determining the fate of its parts.

This revised interpretation, since its appearance in the mid-

1960s, has gained considerable acceptance and support. After more than fifty years of strict avoidance, on behaviorist principles, terms such as "mental imagery" and visual, verbal, auditory "images," and the like have exploded in the last few years into wide usage as explanatory constructs in the literature on cognition, perception, and other higher functions.

The revised interpretation brings the conscious mind into the causal sequence in human decision making—and therefore into behavior generally—and thus back into the realm of experimental science from which it had long been excluded. This swing in psychology and neuroscience away from hard-core materialism and reductionism back toward a new, more acceptable brand of mentalism tends now to restore to the scientific image of human nature much of the dignity, freedom, and other humanistic attributes of which it has long been deprived by the behaviorist approach.

Old metaphysical dualisms and the seemingly irreconcilable paradoxes that formerly prevailed between the realities of inner experience on the one hand and those of experimental brain science on the other become reconciled today in a single comprehensive and unifying view of mind, brain, and man in nature. These changing concepts of mind substantially alter the general image of man and his role as had been drawn in the behaviorist tradition, and also bring other major departures from traditional materialist doctrine.

When subjective values are conceived to have objective consequences in the brain, they no longer need be set off in a realm outside the domain of science. Instead of separating science from values, the present interpretation (when all the various ramifications and logical implications are followed through) leads to a stand in which science becomes the best source, method, and authority for determining the criteria and reference frame for ultimate values and those ultimate ethical axions and guidelines to live and govern by. Under science here, I include broadly the knowledge, understanding, insight, and perspectives that come from science. But more particularly I am thinking of the principles for validity and reliability and credibility of the scientific way as an approach to truth—insofar as the human brain can comprehend truth. In other words, what has disparagingly been called "Scientism" in the past gets a new boost now, with added dimensions and a whole new image and outlook.

On the present terms human values become very much a problem for science, and in certain respects perhaps the most important problem today in the whole of science. As explained earlier, human value priorities stand out as the most strategically powerful causal agents now shaping events on the surface of the globe. More than any other causal system with which science now concerns itself, the human value factor is going to determine the future.

I have rated the problem of human values "Number 1" for science in the coming decade, above the more tangible concerns like poverty, population, energy, or pollution on the following grounds: First, all these conditions are manmade and very largely products of human values. Further, they are not correctable on any long-term basis without first changing the underlying human value priorities involved. And finally, the more strategic way to remedy these global conditions is to go after the social value priorities directly in advance, rather than waiting for the value changes to be forced by worsening conditions. Otherwise we are doomed from here on to live always on the margins of intolerability, for it is not until things get rather intolerable that the voting majority gets around to changing its established values. It is apparent, further, that other approaches to our crisis problems already receive plenty of attention. It is the human value factor that has been selectively neglected and even considered on principle, to be "off limits."

It may be seen that one outcome of all this is that science gets promoted into a somewhat different social role above that of the provision of better things for better living, or the prediction and control of natural phenomena, or even that of advancing knowledge. Science on these terms becomes a major means of helping to shape values and belief systems at the highest level and the most direct avenue to an intimate understanding and rapport with those "forces that move the universe and created man."

5

Bridging Science and Values

Support for science in a role concerned with ultimate value will require a major turn-around in prevailing opinion. The primary task we face throughout is to combat long-accepted views regarding the relation of science to values. This central issue is again confronted here. The reader who already is convinced of the need for change can skip much of this chapter, which reaffirms many points already stated. The present treatment differs in being a more advanced, more detailed and documented discussion with multiple references for the serious scholar, critic, and others who may be interested in the foundations of moral values. The magnitude and social urgency of the issue are taken to warrant the inclusion of this more focused treatment despite the overlap.

General acceptance of the inadequacy of science in the realm of ethical and moral judgment is reflected in the old adage that science deals with facts, not with values, and its corollary that value judgment lie outside the realm of science. Other versions state that science can show us *how* but not *why*, or how to achieve defined goals but cannot tell us which are the morally right or wrong goals to aim for. A further pronouncement holds that science can tell us what *is* but not what *ought* to be, or that science *de*scribes but cannot *pre*scribe.

Although this time-honored dichotomy between science and value judgment has not gone unchallenged (2, 10, 13, 36, 80), the great majority in science, philosophy, and related fields continue today to accept the tradition that science as a discipline must by its very nature deal with objective fact, and that science, either as a method or as a body of knowledge, can neither prescribe values nor resolve issues in the realm of subjective value. When it comes to value conflicts, we are told that we must seek our answers elsewhere—in the humanities, in ethics and philosophy, and particularly in religion, long held to be the prime custodian of human value systems. The basic validity of this traditional separation of science and values and the related limitations it has imposed on the role of science are today open to question in the context of current mind-brain theory.

Value Problems in Scientific Perspective

Human values, in addition to their commonly recognized significance from a personal, religious, or philosophic standpoint, can also be viewed objectively as universal determinants in all human decision making. All decisions boil down to a choice among alternatives of what is most valued, for whatever reasons, and are determined by the particular value system that prevails. From the standpoint of brain function, it is clear that a person's or a society's values directly and constantly shape its actions and decisions. Any given brain will respond differently to the same input and will tend to process the same information into quite diverse behavioral channels depending on its particular system of value priorities. In short, what an individual, or a society, values determines very largely what it does. As human numbers increase and science and technology advance, the regulative control role of the human value factor—which directly determines how all the increased human impact will be applied and directed—becomes correspondingly more powerful.

In a different vein, we are informed that the prevailing social neurosis of our times is a growing sense of valuelessness, apathy, hopelessness, and loss of purpose and higher meaning. We are reminded of the generalized disintegration of long-established values and belief systems, the grasping in all directions for new answers and new life styles, and the reviving in radical form of some of the old answers. From other directions come warnings that the world community needs a whole new system of social value guidelines if civilization is to survive, "new ethics for survival, as Hardin puts it (32)," that would act to preserve our world instead of destroying it.

In view of the enormous timely importance and control power of human values and their critical role in shaping world events, it follows that if science is inherently inadequate by its basic nature to deal with values and value issues, then we are confronted (as the voices of antiscience assert) with what is surely a profound shortcoming in science and all it stands for. On these terms, it is understandable that government should be tightening the screws on the funding of science, especially pure science, and that the overall intellectual confidence in science gradually should be in decline while the forces of antiscience gain new ground through the writings of the critics of science (46, 60). The future of science will be very different depending on whether or not science is recognized in the public mind to have competence in the realm of values. Reciprocally, the future of society also will be very different depending on whether its value perspectives are shaped from the truths and world view of science or from alternative otherworldly determinants that now prevail.

Grounds for Reappraisal

While the separation of science and values seemed to have logical justification in the past, and still does with respect to certain aspects of scientific methodology, new grounds can be seen today that directly challenge the basic philosophic validity of the science-values dichotomy. Recent developments, especially in the behavioral sciences, reopen central issues and argue for a revised philosophy in which modern science becomes the most effective and reliable means available to the human brain for determining credible criteria on which to found moral value and meaning (80). Problems of values, ethics, and morality (questions, that is, of what is good, right, and ethically true and of what ought to be) become, in these revised terms, something to which science can, in the most profound sense, contribute fundamentally and in which science should be actively and responsibly involved.

Although similar proposals since the time of Francis Bacon have been largely written off as scientism by detractors, conceptual developments during the last decade have introduced an interpretation of conscious mind and a resultant philosophical framework that substantially alter the picture. The relation of subjective values to objective science, the scientific status of values, and the kinds of human values supported by science are all directly affected. The current view endowing conscious experience with an active causal role in brain processing directly contradicts the central founding precepts of Watsonian behaviorism and of twentieth-century scientific materialism generally (75, 81). Important departures from long-established determinist and materialist doctrine follow, with extensive implications for the philosophy of science and the derivation of values.

The involved theoretical changes have been outlined extensively elsewhere (73–83) and may be reviewed briefly as follows: we reject prior theories of consciousness which interpret subjective experience to be an epiphenomenon; an inner aspect or any form of passive, parallelist correlate of brain activity; or identical to neural events as in "psychophysical identity theory." Conscious phenomena in this revised model are "different from, more than, and not reducible to" neural events—though built of neural and perhaps glial and other physicochemical events as elements. We also reject the view that consciousness is a pseudoproblem conjured into our thinking as an artifact of semantics that can be resolved with a proper linguistic approach. Bypassing all the foregoing, the theory centers on the interpretation of consciousness as an emergent property of brain activity, as upheld especially by the Gestalt school of psychology in ideas that peaked during the 1930s and early 1940s (7, 37, 38, 40).

The present model is distinguished from the earlier emergent Gestalt concepts in that, first, the emergent properties are not conceived to be correlated with, nor derived from cortical electric field or volume current conduction effects, but are conceived rather in terms of traditional nerve-circuit and cerebral-integration theory. Second, the present model does not require isomorphic or topological correspondence between the emergent subjective properties and the neural events. Subjective meaning is conceived to derive rather from the functional or operational impact or the way a given brain process "works" in the context of brain dynamics (67).

Third, while agreeing with Gestalt theory that conscious phenomena are not reducible to neural events, the current view does not take the extreme Gestalt position that categorically rejects analysis and explanation in terms of the parts. In the present view, a description of the neural events generating any conscious experience would have tremendous explanatory value and would seem to constitute the best hope for advancing understanding. Fourth, and most important, the emergent properties in the present view are not interpreted to be mere passive parallel correlates, aspects, or by-products of cortical events, but rather to be active causal determinants essential to the control of normal cerebral processing.

A conceptual explanatory model is provided for the way mind can rule matter in the brain and exert causal influence in the guidance and control of behavior, on terms acceptable to neuroscience and without violating monistic principles of scientific explanation. No direct empirical proof is available, of course, but neither is there proof available for the traditional behaviorist-materialist position. It comes down to a balance in credibility, all things considered, and one can only say that many of us have come increasingly during the past decade to regard this modified causal concept of conscious mind as being more credible on several counts than the behaviorist view.

No dualistic interaction in the classical sense is implied. The causal power attributed to the subjective properties resides in the hierarchical organization of the nervous system and in the power exerted by any whole over its parts. Mind moves matter in the brain in much the same way that an organism moves its component organs and cells, or a molecule governs the travel course of its own atoms, electrons, and subnuclear elements in a chemical reaction. In the case of conscious experience, it is the dynamic system properties of high-order cerebral processes that control their component neural and chemical elements. Holistic, emergent, or system properties elsewhere are recognized universally to have causal efficacy. The present view merely asserts that the emergent subjective properties of brain processing are no exception to the general rule. The principles of interlevel hierarchic causal control involved here have since been amplified and expounded upon in some detail by Pols in a philosophical framework (52, 53).

The present view differs from earlier concepts in that the brain-behavior sciences can no longer ignore subjective conscious experience and expect to obtain, in principle, a complete objective description of higher, psychological functions. The conscious properties, per se, are conceived to make a profound difference in the course of neural events. Subjective experience in the present scheme is put to work in brain function and given a reason for being and for having evolved in a physical system. Stemming largely from efforts to account for the inferred unity and/or duality of subjective awareness in the presence or absence of the cerebral commissures (74, 85), this modified approach to the mind-brain interface brings together selected features of prior materialist, mentalist, emergent, and pragmatist doctrines in a novel combination. The result, in effect, means that the whole value-rich, qualitative world of inner, conscious, subjective experience (the world of the humanities) which has long been explicitly excluded from the domain of science on behaviorist-materialist principles, becomes reinstated.

The science-values dichotomy is directly affected in two ways: Subjective values are in principle no longer excluded from the realm of experimental science and scientific method, and, secondly, the world view of science, and hence the kinds of human values supported by science, are greatly altered in humanistic quality. These two factors, taken singly and in combination with their ramifications and implications, revise and directly counter prior arguments for keeping "value judgments outside the realm of science."

Revised Outlook

The involved change in the scientific status of conscious mind carries with it a renunciation of much of the mechanistic, behavioristic, deterministic, and reductionist thinking that formerly had characterized science and which the humanities have always found difficult to accept. Behavioral science in particular acquires a new image in this respect and becomes much more subjective and humanistic. Recent trends in psychology, referred to variously as the "humanist," "cognitive," or "third" revolution, or simply as the "new psychology," are more than just a matter of changing attitudes in science, material progress, or passing social trends. They have authentic theoretical grounding in fundamental changes in our basic mind-brain concepts.

On the above terms it becomes increasingly impossible, among other things, to accept the idea of two separate realms of knowledge, existence, or truth: one for objective science and another for subjective experience and values. Old metaphysical dualisms and the seemingly irreconcilable paradoxes that have prevailed in psychology (88) between the realities of inner experience on the one hand and those of experimental brain research on the other disappear into a single, continuous hierarchy. Within the brain, we pass conceptually in a hierarchical continuum from the brain's subnuclear particles, on up through the atoms, molecules, and brain cells to the level of nerve-circuit systems without consciousness, and finally to cerebral processes with consciousness. Objective facts and subjective values become parts of the same universe of discourse. The hiatus between science and values is erased in part by expanding the scope of science to encompass inner experience and by altering the status of subjective values so that they are no longer set off in an epiphenomenal or other parallelistic domain outside the reach of science.

So long as science disclaims and rejects the entire realm of inner subjective experience as being acausal, the content and world view of science remain inadequate and unsatisfying for answers that involve subjective value. With the acceptance of a causal concept of conscious experience, the qualitative, subjective dimensions in value systems no longer exclude a scientific approach; nor are these subjective dimensions necessarily neglected or demeaned. The scientific image of man regains much of the freedom, dignity, and other humanistic attributes of which it has long been deprived. Many prior antiscience objections to the mixing of science and values no longer apply. A holistic world model and interpretation of reality is also supported, in which the qualitative pattern properties of all entities are conceived to be just as real and causally potent as are the properties of their elements or their quantitative measurements and abstractions. This preservation of the qualitative value and pluralistic richness of physical reality stands counter to the common tendency to correlate science with reductionism (60).

Further Humanist Implications

A substantially altered picture of causal determinism in behavior is now inferred in which all subjective mental phenomena, including subjective values, are recognized to have a causal role per se in the decision-making process, rather than being mere correlates or aspects of a self-sufficient brain physiology (71–83). In any decision to act, the conscious mental phenomena override and supersede the component physiological and biochemical determinants. Even subjective feelings about projected outcomes anticipated to result from a given choice as far as twenty-five or a hundred years in the future may be entered proactively as causal determinants in the cerebral operations that lead to a given choice. Behavior on these terms is still causal and deterministic but at a cognitive and conative, mental (rather than mechanistic or physiological) level. Total freedom from causation would result in meaningless, random chaos and would be as bad or worse than mechanistic determinism. Current theory provides a compromise that allows one to determine one's own actions according to one's own subjective wants, personal judgment, perspectives, cognitive aims, emotional biases, and other mental inclinations. The degree and kinds of freedom of choice introduced thereby into the causal sequence of decision making clearly set the human brain apart, at an apex post in the universe above all other known systems, in its ability to choose and control a course of events.

The concepts raised in the foregoing sections are central and basic to ethics and value theory. Value priorities, especially in the ideological, religious, and cultural areas, are heavily dependent, directly or by implication, on concepts and beliefs regarding the properties of conscious mind and the kinds of life goals and world views which they allow. Directly and indirectly, social values depend on whether consciousness is believed to be mortal, or immortal, reincarnate, or cosmic, and whether consciousness is conceived to be localized and brain-bound, or universal in essence—as in panpsychism or Whiteheadian theory—or perhaps capable of "supracoalescence" in a megamind. Where formerly there were seemingly unlimited degrees of freedom for speculation in these areas, advances in neuroscience continue to narrow the latitudes for possible realistic answers. In modern neurophysiology, it is not so much a question of whether conscious experience is tied to the living brain, but rather to what particular parts of the brain, or to which neural systems and under what physiological conditions (16, 22, 41).

As the brain process comes to be understood objectively, all mental phenomena, including the generation of values, can be treated as causal agents in human decision making. The origins, directive potency, and consequences of values all become subject, in principle, to objective scientific investigation and analysis. This applies at all levels, from that of the brain's pleasure-pain centers and other reinforcement systems on up through the psychosocial, economic, and related forces that mold priorities at the societal, national, and international planes. Modern behavioral science already treats value variables and their formation as important causal variants in behavior, and it also deals analytically with goals, needs, motivation, and related factors at individual, group, and societal levels. What amounts to a science of values in the context of decision theory becomes conceivable (2, 58, 80), extending into all branches of behavioral science and forming a skeletal core for social and behavioral science. For neuroscience it suggests a design principle for understanding brain organization and cerebral processing as a goal-directed, value-guided decision system, replacing older "stimulus-response" and "central switchboard" concepts that arose out of spinal cord physiology.

"Is"–"Ought" Fallacy

On the foregoing terms, most remaining traditional objections to the mixing of science and values tend to disappear. Probably the most influential factor currently sustaining the science-values dichotomy is the prevailing acceptance of the contention of professional philosophy that it is logically impossible to determine what "ought to be" from what "is," or to derive ethical priorities from objective facts. I think this oft-cited dictum has never been defensible from the standpoint of behavioral science and is best appraised as a logical artifact of a strictly pencil-and-paper approach in philosophy. Human values are inherently properties of brain activity, and we invite logical confusion by trying to treat them as if they had an independent existence artifically separated from the functioning brain. In the operations of the brain, incoming facts regularly interact with and shape values.

These two factors, "internal" and "external," interact as cofunctions in the building of one's sense of values. The resultant value system of any adult or society, along with related concepts of what ought to be, is determined in very large part by the body of facts encountered. In terms of cerebral processing, it is difficult to see a better way to determine what ought to be than on the basis of factual information, especially facts and deductions which have been scientifically verified. History and common observation confirm that nothing is more proficient than science at prescribing what ought to prevail in order to achieve almost any defined aim, whether this be a landing site on Mars, improved physical or mental health, or whatever. The same applies in regard to ultimate aims as discussed below under "Prime Determinants."

In the processing of factual input, the brain mechanisms already are richly equipped in advance with established value determinants and intrinsic logical constraints in the form of combined innate and acquired needs, aims, and motivational and other goal-directed factors that have their origins partly in biological heritage and partly in prior experience, and that may also come formally through rational acceptance of ethical axioms. Particularly pertinent to this processing is the largely inherent need of the human brain to try to perceive meaning, including that for its "self" in the long term. Since, in practice, it is never a matter of deriving values from extrinsic facts per se, the question at issue may be framed more properly in terms of the impact of a set of facts upon ongoing brain processes. Accordingly, if one asks whether a set of facts can shape value priorities or one's judgment of what ought to be, the answer, of course, is yes. We are constantly adjusting our ethical values to conform with new factual information, and the advance of science historically has always had a deep, inevitable influence on human value-belief systems.

For present purposes, the innate primal system of values that is based on biological survival (58) and is part of human nature, the personal, interpersonal, and "humanitarian" aspects of which tend to form a large common denominator for all ethical systems, is treated here largely as a constant. The present focus is on areas where ethical systems disagree and particularly on those cognitive, axiological, and related variables that are affected by acceptance or rejection of the method and world view of science as a final frame of reference. It is in this cognitive, rational realm that most major value conflicts and ideological differences are found. The further, related problem of the prime determinants and starting axioms and premises in ethical systems is considered separately below.

Convergent Reasoning

Other convergent lines of reasoning can be seen that support the same conclusions reached here regarding the potential role of science in shaping values, so that if the current mind-brain model is shot down, most of these conclusions still hold on other grounds. Common sense dictates that science, as man's number one source of factual information, should be enlisted in the realm of value judgment on the simple rationale that an informed judgment is generally preferable to one that is uninformed or misinformed. Similarly, if moral judgments about right and wrong are best arrived at on the basis of what is true, avoiding what is false, science would seem on this count as well to deserve a leading role in shaping ethical values instead of being disqualified. The word science throughout this chapter does not refer to individual scientists or their personal opinions and values but rather to the total collective knowledge and world model drawn from all the sciences, including the social and political; and to the insight, understanding, and sense of value fostered by this total collective (the nearest thing to omniscience available to human society). The overall perspectives of science, taken in this broad empirical sense, may often be better reflected in the thinking lay citizen than in the scientific specialist. The reference to science is also a reference to the relative validity, credibility, and reliability of the scientific method itself as an avenue to belief and an approach to truth so far as the human brain can know it. The main dichotomy involved is between a monistic, this-world conception and dualistic, other-worldly schemes for our cosmos and reality.

A quite different argument renounces any active approach to ethics through science, not because social values should be left to the humanities, the Church, or to Marx, but rather on grounds that it is wiser that values be left to themselves to change spontaneously, by collective intuition as it were, in response to changing environmental conditions. Some economic realists assert that this is the only way that values change, and they eschew any moral philosophizing or prescriptive idealizing, viewing it as ineffectual. This stance overlooks the strong reciprocal interaction between mental concepts and environmental conditions, and the tremendous impact that ideology and value systems have always had on the course of human history. It also overlooks the fact that social values formed on this situational feedback basis as a reflection of prevailing conditions tend to be locked in a democratic society to levels of bare tolerability rather than to ideal optimums, for reasons already mentioned.

It is not only the value systems of orthodox religion that have been found wanting today, but also those based on humanist, communist, existential, and even common humanitarian principles. Recourse to recent alternatives like the lifeboat ethic or the battlefront ethic of triage, as currently formulated, hardly offers inspired solutions. Current world conditions call for a unified global approach with value perspectives built on something higher than just the human species or its societal dynamics, something more godlike that will include the welfare of the total biosphere and ecosystem as a whole on an evolutionary time scale. The greater the human impact on the ecosystem, the more urgently these higher perspectives are needed. They are imperative also in efforts to perceive higher meaning, where it becomes a logical necessity that humanity be able to perceive itself in terms of a meaningful relation to something more important than itself.

The Value Hierarchy and Prime Determinants

The more critical value issues that must be faced in the near future will involve decisions that ultimately require appraisals of the relative worth of human life in various contexts. For example, as world crowding conditions get tighter, the value of human life must be balanced increasingly against that of other species. Having already destroyed the natural meaning and dignity of life for a number of subordinate species and permanently extinguished others, man will be forced to judge how much further species' rights can be ignored and by what ethic. Many more examples can be listed in which scientific advancements, coupled with mounting population and related pressures, have raised a growing host of moral dilemmas that revolve finally around the question of the ultimate worth of life itself (15). Possible answers become relative with alternatives that call for assessment within some larger ethic yet to be found. What is needed ideally, of course, to make decisions in these areas is a consensus on some supreme comprehension and interpretation of the universe and the place and role within it of man and the life experience.

The same position is reached by way of abstract value theory, in which it is shown that values depend largely on goals, and that any concept or belief regarding the goal and value of life as a whole, once accepted, then logically supersedes and conditions the entire hierarchy of value priorities at all subsidiary levels (80). Values at the ideological plane become ordered, and ethical issues judged in accordance with the conceived ultimate goal and purpose of life as a whole. This latter will logically imply, in turn, an associated world view or universe scheme that is consistent.

By one route or another we come down to these prime determinants of value priorities—these life goal, world model concepts and beliefs, explicit or implied, that lie at the heart of the problem of moral judgment and pose the central challenge. This is where the great unknowns lie, and also where the great differences of opinion are found. This is where answers are most needed, and where any answers, right or wrong, once accepted, have the greatest impact. And it is here also that the competence of science in the arena of values, and any new ethic must eventually be proven. The scientist, trained to rigorous reasoning and skepticism, to checking against detailed empirical evidence, and above all to avoiding false conclusions, may easily be persuaded at this point that value problems are not for science. Let us be reminded, however, that final, absolute, or perfect answers are not demanded, only improved ones, and that society has in the past and probably will continue in the future to find and abide by some kind of answers from somewhere. The question is not whether science can provide final, complete, or perfect answers, but whether there is any alternative that does as well by long-term, "future generation" standards.

Changing to an ethic based in the truths of science would entail in large part a substitution of the natural cosmos of science for the different mythological, intuitive, mystical, or otherworldly frames of reference by which man has variously tried to live and find meaning. World view concepts setting the parameters for higher meaning would need to be reinterpreted in terms of the insights of science, and the consequent value-belief implications analyzed and formulated. A similar approach and set of reformulations is arrived at from other sources by Ralph Burhoe (11), who describes the effort as "scientific theology." Again, it must be remembered throughout that we are talking of nonreductive science in the new emergent *mentalist* or holist paradigm as explained in further detail in chapter 6.

The result, predictably, would not much alter the large majority of present, day-to-day values nor many of the traditional moral and ethical teachings concerning personal and interpersonal conduct upheld in belief systems in the past that have proven themselves through history. At the same time, significant changes would be expected in areas more directly dependent on world view perspectives. An increased respect and reverence for evolving nature along with its endless wonder and beauty and for what is sometimes referred to as the "infinite wisdom of nature" may be directly inferred along with added concern for the balance, progressive differentiation, and quality of the ecosystem as a whole. Emphasis must be placed on the term "evolving," since the bad things in nature are as natural as the good. It is in the *trends* of the creative process toward improved quality of existence that one is enabled to perceive right-wrong differences.

However, this is not the place to attempt to undertake the enormous task of trying to further analyze and define the particular kinds of social value changes that might be incurred in the adoption of a this-world ethic based on the validity and world view of science. The present aim is only to help justify and clear the way by trying to remove a first major hurdle. Once it can be shown to be intellectually respectable to use and apply the facts and views of this-world reality, as revealed in science, to the realm of value judgment, thinking along these lines will begin to develop on many fronts.

Looking ahead, it may be remembered in this connection that social decision making does not require, and frequently does not involve nor wait upon, precise logical answers or directives, but proceeds on the basis of vague impressions, personal biases, emotional leanings, general attitudes, and the like. This is why any change that can be achieved merely in the general public attitude regarding the relation of science to values and higher meaning, that may help to counter antiscience and reductionist fallacies, or the obsolete tradition that science and values do not mix—even though it remains only as a vague impression in the minds of most of the voting majority—would nevertheless have an enormous consequence on decision making. The pervasive impact via a vast complex of social decisions on things like population policy, global conservation, and related ecosystem planning generally, could add up to an overall potential future benefit far exceeding that of many other top scientific goals—such as conquering cancer or schizophrenia—especially from the standpoint of coming generations.

6

Mind-Brain Interaction

Mentalism, Yes; Dualism, No

Are the things we hold most sacred, including the human psyche and forces of creation, best conceived in dualistic, otherwordly terms? Historically tied to interpretations of the conscious mind, the case for dualist forms of existence was effectively countered and held down by materialist science to a negligible status for many decades. Dualist concepts made a notable comeback during the 1970s, however, and again today receive vigorous support from some authorities as a viable answer to the mind-brain problem. The present critique, addressed largely to specialists, questions the logical foundations of the new dualist position.

The New Interactionist Philosophy

When two eminent authorities of science and philosophy, of the stature and influence of Sir John Eccles and Sir Karl Popper, join forces to affirm dualistic beliefs in the reality of the supernatural and the existence of extraphysical, unembodied agents to challenge some of the most fundamental precepts of science, one is impelled to take more than passing notice. Regardless of one's personal convictions and reactions, the kind of public message that is conveyed, directly and indirectly, by their book *The Self and Its Brain: An Argument for Interactionism* (1977) along with Eccles' more recent volume *The Human Mystery* (1979), and the potential impact of these on the intellectual perspectives of our times become a matter of some concern. Such considerations, and the fact that my own views and writings are cited in support of some of the key concepts and as being in alignment with dualist interactionism, prompt this effort to clarify certain points that otherwise leave erroneous impressions.

Before I attempt to focus on specifics, it will help to mention broadly that whereas Sir John Eccles and I have similar outlooks with many highly congenial perspectives, aims, and values, we do, however, share certain friendly differences in regard to the nature and locus of consciousness and the support of dualism. I have always favored monism, and still do. Sir John tells me that I am a dualist and I respond: Only if the term is redefined to take on a new meaning quite different from what it traditionally has stood for in philosophy. Dualism and monism have long represented a dichotomy that offers opposing answers to one of man's most critical and enduring concerns, namely, can conscious experience exist apart from the brain? Dualism, affirming the existence of independent mental and physical worlds, says yes and opens the door to a conscious afterlife and to many kinds of supernatural, paranormal, and otherworldly beliefs. Monism, on the other hand, restricts its answers to one-world dimensions and says no to an independent existence of conscious mind apart from the functioning brain.

In recent years there has arisen some real need to change and sharpen definitions of certain philosophic terms to fit our new views in neuroscience. However, in the case of monism and dualism, I see no advantage in changing the classic definitions. We greatly need terms by which to distinguish the critical dichotomy regarding the potential separability of brain and conscious experience during life as well as after. Dualism and monism have long served this need in the past and seem best qualified to continue.

At the same time I am in strong agreement with Eccles in rejecting both materialism (or physicalism) and reductionism—or at least what these terms predominantly stood for prior to the mid-sixties. Since 1965 I have referred to myself as a mentalist and since the mid-thirties have firmly renounced reductionism in the philosophic, "nothing but" sense to be explained below. However, in the case of the terms mentalism and the opposing materialism, and the form of dichotomy these two imply, some change and sharpening of definitions is now called for by our modified mind-brain concepts. On our new terms, which I will outline below, mentalism is no longer synonymous with dualism nor is physicalism the equivalent of monism. By our current mind-brain theory, monism has to include subjective mental properties as causal realities. This is not the case with physicalism or materialism which are the understood antitheses of mentalism, and have traditionally excluded mental phenomena as causal realities. In calling myself a mentalist, I hold subjective mental phenomena to be primary, causally potent realities as they are experienced subjectively, different from, more than, and not reducible to their physicochemical elements. At the same time, I define this position and the mind-brain theory on which it is based as monistic and see it as a major deterrent to dualism. To better explain these distinctions, it will be helpful to start at the beginning and to follow the conceptual developments step by step as they occurred.

Conceptual Breakthrough

My long-trusted materialist logic was first shaken in the spring of 1964 in preparing a nontechnical lecture on brain evolution in which I was extending the concept of emergent control of higher over lower forces in nested hierarchies to include the mind-brain relation. I found myself concluding with the then awkward notion that emergent mental powers must logically exert downward causal control over electrophysiological events in brain activity. Mental forces were inferred to be equally or more potent in brain dynamics than are the forces operating at the cellular, molecular and atomic levels (71). Again, in September of that year, when preparing a paper for the Vatican Conference on Brain and Consciousness organized by John Eccles, it occurred to me that the functionist interpretation of consciousness that I had outlined in the early fifties (67) and still favor, must also logically call for a functional (and therefore *causal*) influence of conscious experience in brain activity. It was obvious that these combined concepts, were they to hold up, would provide a new approach to the old question of how consciousness may be of functional use and exert a causal control role in brain processing. The kind of psychophysical relation envisaged showed how mind could influence matter in the brain, making the interaction of such different things as mental states and physical events logically understandable at long last on terms that were scientifically acceptable.

In the mid-sixties, such interactionist concepts were still complete heresy in neuroscience and I did not venture to push them at the Vatican conference beyond mild reference to "a view that holds that consciousness may have some operational and causal use." To this Eccles responded by asking: "Why do we have to be conscious at all? We can, in principle, explain all our input-output performance in terms of activity of neuronal circuits; and, consequently, consciousness seems to be absolutely unnecessary!" (19) This is, of course, what we had all been taught and believed for decades, not only in science but also (by the great majority) in philosophy. The idea that the objective physical brain process is causally complete in itself without reference to conscious or mental forces represents the central premise of behaviorism and of scientific materialism in general and has long served as a prime basis for the renunciation of the phenomena of subjective experience as explanatory constructs in science. Eccles, however, already at the time a dualist by faith, training, and publication (18), went on to add: "I don't believe this story, of course; but at the same time, I do not know the logical answer to it." Nevertheless, his considered conviction on the first point was firmly reiterated in a later session: "I am prepared to say that as neurophysiologists we simply have no use for consciousness in our attempts to explain how the nervous system works" (Eccles 1966).

I argued the point briefly but was not yet sufficiently versed in my new-found answer to pursue it vigorously at the time. In the ensuing weeks and months, however, in pondering the unifying role of callosal activity, the ideas kept recurring and the more I thought about them, the better they looked. A trial run the next April to our Caltech Division of Biology convinced me that reductive neuroscience and biology were not exactly ready for this kind of thinking. However, I decided to proceed anyway with a presentation the following month in a humanist lecture at the

University of Chicago for the volume, *New Views of the Nature of Man*, edited by John Platt. For the purpose of this lecture, I worked the new mind-brain ideas into a discussion of holist-reductionist issues, emergent downward control, and "nothing but" fallacies in human value systems, in a broad refutation of the then prevalent "mechanistic, materialistic, behavioristic, fatalistic, reductionistic view of the nature of mind and psyche." It was on this occasion that I openly changed my alignment from behaviorist materialism to antimechanistic and nonreductive mentalism (as the term mentalism is used in psychology in contrast to behaviorism; not, of course, in the extreme philosophic sense that would deny material reality). At the same time, I described this new position as a unifying scheme that "would eliminate the old dualistic confusions" in favor of "a single this-world measuring stick for evaluating man and existence."

Mind Moves Matter in the Brain

The main thesis of the essay, as in the Popper and Eccles book, was psychophysical interaction, its logical support and its scientific, philosophic, and human value implications. Essentially, it presented the view that subjective experience as an operational derivative and emergent property of brain activity plays a prime causal role in the control of brain function. It differed from previous emergent theories of consciousness, from C. Lloyd Morgan (47) onward, in that earlier emergent views of mind had been conceived in terms that were parallelistic, double aspect, or epiphenomenal, and had rejected any direct causal influence of mental qualities on neural processing (39). The thesis was focused on contradicting the traditional, mechanistic assumption expressed by Eccles that brain processing can be completely accounted for, in principle, without including conscious phenomena. Presented in terms of neuronal circuitry and concepts of neuroscience, my theory seemed to counter and refute for the first time on its own grounds, the classic physicalist assumption of a purely physical determinacy of the central nervous system. Subjective mental phenomena had to be included. Mind-brain interaction was made a scientifically tenable and even plausible concept without reducing the qualitative richness of mental properties. The overall aim of the paper, as in the Popper and Eccles volume, was to show that this recognition of the primacy of conscious mind as causal would alter profoundly the value implications of science which were being downgraded by the then strongly dominant philosophy of reductive mechanistic materialism.

At the same time, the proposed mind-brain model was taken to undermine dualism as well by explaining conscious experience in terms that would make mind inextricably inseparable from, and embodied in, the functioning brain. It provided a rationale for the evolution of mind from matter and also for the emergence of mind from matter in brain development. Presented as a "conceptual skeleton on which to build a body of philosophy," I described it as a scheme that "would put mind back into the brain of objective science and in a position of top command."

When the reprints arrived, I sent my new mind-brain "answer" to Eccles, who previously had expressed little, if any, active interest in the holist-reductionist issues (19). I was delighted to see by his next International Brain Research Organization presentation (20) that he had clearly joined our ranks as an ardent antireductionist denouncing "the materialistic, mechanistic, behavioristic, and cybernetic concepts of man." Reversing his earlier stand on the uselessness of consciousness for a full account of brain function, Eccles has since lent his support to the new logic for the causal influence of mind over neural activity. On these points I believe we have remained in good general agreement (21).

It is in regard to the nature of the causal influence and to the use of these new mind-brain concepts to support dualism that our critical differences arise. Other differences concerning the relation of consciousness to the right hemisphere, to language, to animals, and to self-consciousness, though of some concern, are of secondary importance in the present context and will not be pursued here. However, in the case of those differences that pertain to the mind-brain problem and to dualist interactionism we deal, as Eccles has very ably emphasized, with more than ordinary professional and academic interpretations. At stake are central key concepts that directly involve fundamental convictions regarding the nature of man's inner being, physical reality, the meaning of existence, and related matters of ultimate concern. Because perspectives in this area profoundly shape human value systems and societal decision making and hence human destiny, we mutually agree that these issues must take precedence over other considerations.

Looking back today, it seems clear that I quite failed to foresee how the new mind-brain solution might be taken to support dualism. Even though dualism and mentalism had long been associated and even equated, and some colleagues had forewarned that I might accordingly be accused of dualism, I nevertheless supposed the new mentalist-dualist distinctions to have been sufficiently clarified (77). Back in the 1960s dualist views were no threat to science and accordingly, it seemed much more important in those years to combat the more prevalent errors of materialism, mechanism, behaviorism, and reductionism, than to emphasize the conjoint logic against dualism. Again, the finer points involved here are better and more easily explained if we continue to follow the chronological approach.

Growing Scientific Acceptance

After waiting more than three years during which the feedback was mostly positive, especially from humanist groups, I tested the theory more directly in the scientific community by presenting it at a neurological meeting (78) and then to the National Academy of Sciences (76). A follow up article derived from my talks appeared in the *Psychological Review* (75) and was reprinted several times. The result was a wide exposure, including a critique (5) and my reply (77) to it, within those disciplines most knowledgeable and most apt to be critical. In these conjectural areas where the concepts are still beyond any direct experimental verification, the next best test is to put them in the marketplace to be churned over by hundreds of minds from all different angles. In this respect the years 1969 to 1971 were the critical years for this theory. No logical flaw nor prior statement, so far as I know, has yet come to light.

By the early 1970s, the modified concept of consciousness as having causal efficacy began to gain substantial scientific acceptance, particularly in psychology in a pervasive resurgence of mentalism and antibehaviorism that is still gathering momentum (35). Essentially, the new interpretation brought a logical change in the scientific status of subjective experience, replacing behaviorist principles with a mentalist or cognitivist paradigm. Psychologists could now refute the logic and principles of behaviorism and refer directly to the causal influence of mental images, ideas, inner feelings, and other subjective phenomena as explanatory constructs. The suddenness with which this began to occur was almost explosive in the cognitive disciplines (59). The movement has already been referred to as the "cognitive revolution" (14) and also variously as the "humanist", "consciousness," or "third" revolution (45), and has extended also into philosophy, anthropology (28), and neuroscience (9, 33, 43).

Eccles' increasingly vigorous campaign for dualist interactionism during this same period has followed on a curve that closely parallels the above. A similar curve can be drawn for a rising public belief in psychic, paranormal, and related mentalist phenomena, along with mysticism, occultism, and other dualist beliefs in the supernatural and in otherworldly forms of existence. Some of these have logical support in the new mind-brain concepts; others are bolstered only spuriously by association. There is good reason to think that the gains made by these mentalist-related developments during this period have been substantially aided, directly and indirectly, by the appearance in neuroscience of a plausible, logical answer by which to counter the basic premises and principles of the traditional materialist paradigm. Without a convincing alternative to replace the physicalist logic, we would be back today much where we were in the mid-sixties, i.e., where materialist-behaviorist reasoning effectively outweighed all the intuitive, natural, and omnipresent subjectivist pressures and arguments, and where cognitive psychology remained in principle a science of para- and epiphenomena. More specificially, the increasing assurance with which Eccles has been able in recent years openly to proclaim dualist arguments not visible in the 1964 conference suggests that he has developed in the interim a new "logical answer" that was not perceived earlier.

How Many New Mind-Brain Solutions?

A first question that needs to be considered is whether the set of concepts which Eccles currently uses to support dualism (Karl Popper's arguments will be discussed separately) is significantly different from that which I proposed as an anti-dualist, monist solution. Have we independently come on two different answers for mind-brain interaction, or is it a matter of different interpretations of basically the same solution? So far as I am able to determine, the underlying concepts by which psychophysical interaction is inferred by Eccles do not differ in any relevant respect from those which I have presented as mentalist monism. In searching the arguments and evidence advanced by Eccles (57) one finds much the same reasoning that I have used to support my own concept of consciousness (67–78). The phraseology and emphasis are somewhat different, and some different neural examples of the principles are introduced, but the conceptual model for mind-brain interaction that is inferred seems entirely consistent and certainly no distinct alternative is offered.

Eccles emphasizes with italics (p. 362) that "a key component of the hypothesis is that unity of conscious experience is provided by the mind and not by the neural machinery," and this point is again stressed in Dialogue VIII, p. 512, and again in his Gifford Lectures (23). Here we are in full accord. I too had made precisely the same point in 1952, stating: "In the scheme proposed here, it is contended that unity in subjective experience does not derive from any kind of parallel unity in the brain processes. Conscious unity is conceived rather as a functional or operational derivative," and "there need be little or nothing of a unitary nature about the physiological processes themselves." In his earlier thinking Eccles had given priority to quite a different concept, expressed in terms of extraphysical "ghostly influences" affecting the course of synaptic events (18). I have since referred to and consistently reiterated the above explanation of mental unity in reference to the role of the cerebral commissures and to the "graininess" problem, emphasizing that the subjective unity does not correlate with the array of excitatory details comprising the infrastructure of the brain process but rather with the holistic mental properties (72–86).

In a reflective appraisal near the end of their volume (57), Eccles observes, "As we have developed our hypothesis, we have returned to the views of past philosophies that the mental phenomena are now ascendant again over the material phenomena." Similarly, I too from the start have described the hypothesis as one that "puts mind back over matter" (73) and "would restore mind to its old prestigious position over matter" (78). That our key concepts for this and for mind-brain interaction in general are essentially one and the same is further indicated where Eccles (57) ends the condensed summary of his hypothesis (p. 373) with the statement: "Sperry has made a similar proposal (Sperry 1969)" and in another "very brief summary or outline of the theory" (p. 495) concludes: "Thus, in agreement with Sperry, it is postulated that the self-conscious mind exercises a superior interpretative and controlling role upon the neural events."

Insights of Karl Popper

When we turn to the solution of the mind-brain problem upheld by Sir Karl Popper, we find it is also basically the same, but the history of its acquisition is quite different. Prior to 1965, Popper's support of dualism rested mainly on the argument that no causal physical theory of the descriptive, argumentative functions, of language is possible. Products of the mind, like myths, abstractions, and mathematical formulas cannot be accounted for by the laws of physiology or physics (54). During the years in which this argument was propounded, it failed by itself to have much influence in countering physicalist objections that products of the mind have neural correlates and that the products of the mind, like other mental entities, were better interpreted in parallelistic terms as being epiphenomena, inner aspects of, or identical to their neurological correlates. As expressed by Oppenheimer and Putnam (48):

> **It is not sufficient, for example, simply to advance the claim that certain phenomena considered to be specifically human, such as the use of verbal language in an abstract and generalized way, can never be explained on the basis of neurophysiological theory, or to make the claim that this conceptual capacity distinguishes man in principle and not only in degree from nonhuman animals.**

In 1965, Popper proposed a new solution to the mind-brain relation that was exactly what he had been looking for in his earlier arguments and which has since become a major theme of his philosophy (56). In a lecture devoted first to a discussion of physical indeterminism, and in a departure from his prior long-time concerns with the logic of knowing, Popper (55) added a second theme concerning some revised perspectives on evolution which he then extended to include the body-mind problem. He emerged with what seems to be basically the same view of evolution and the mind-brain relation that I too had proposed a year earlier in a James Arthur Lecture. In essence, the idea of emerging hierarchic controls is applied to the mind-brain relation. This 1965 switch in Popper's philosophy from a position in which evolutionary theory was held to be tautological, explaining almost nothing, to one in which it explains almost everything was offered with "many apologies," as a development for which he was obliged "to eat humble pie." In line with the main theme of his lecture, a plastic indeterminacy of the emergent controls was emphasized, but the degree of looseness or tightness in the controls is not a critical part of the argument.

Because these concepts concerning hierarchic organization and downward control are crucial both to the Popper and Eccles volume and to the present chapter, I restate them with exact quotes:

> **Evolution keeps complicating the universe by adding new phenomena that have new properties and new forces and that are regulated by new scientific principles and new scientific laws—all for future scientists in their respective disciplines to discover and formulate. Note also that the old simple laws and primeval forces of the hydrogen age never get lost or canceled in the process of compounding the**

compounds. They do, however, get superseded, overwhelmed, and outclassed by the higher-level forces as these successively appear at the atomic, the molecular and the cellular and higher levels . . .

. . . recall that a molecule in many respects is the master of its inner atoms and electrons. The latter are hauled and forced about in chemical interactions by the overall configurational properties of the whole molecule. At the same time, if our given molecule is itself part of a single-celled organism such as paramecium, it in turn is obliged, with all its parts and its partners, to follow along a trail of events in time and space determined largely by the extrinsic overall dynamics of *Paramecium caudatum*. When it comes to brains, remember that the simpler electric, atomic, molecular, and cellular forces and laws, though still present and operating, have been superseded by the configurational forces of higher-level mechanisms. At the top, in the human brain, these include the powers of perception, cognition, reason, judgment, and the like, the operational, causal effects and forces of which are equally or more potent in brain dynamics than are the outclassed inner chemical forces. (71).

Note that this statement includes the basic key concepts on which the Popper and Eccles case for mind-brain interaction mainly rests, i.e., the downard causal control influence of higher emergent (mental) over lower (neural) entities, and the fact that the mental and neural events are different kinds of phenomena regulated by different kinds of laws and forces.

Hence, from very different backgrounds, Popper and I had arrived by 1965 at the same answer to Eccles' problem. Popper presented his as an answer to "a new view of evolution" and "a different view of the world." I presented mine as "a scientific theory of mind" and "a long-sought unifying view of man in nature." We both offered our view as a new solution to the mind-body problem. When one considers that this new turn in Popper's thinking had not appeared in his extensive philosophical publications over the previous forty years, the timing of these convergent developments is remarkable.

In Popper's case, his new solution did not become generally available apparently, except by offprint request, until the lecture he gave in 1965 came to be published in 1972 among other philosophic essays in the volume *Objective Knowledge*. Even Popper's own thinking seems curiously to have been little influenced during this interim. His long article "On the Theory of the Objective

Mind," prepared for the 1972 volume out of two previous papers from 1968 and 1970, introduces his "three world" terminology. It deals with a subject that, unlike the 1965 lecture, almost cries for the use and application of the new mind-brain solution and different view of the world, yet this goes unmentioned. Even in his subsection on the causal relations between the three worlds, he does not refer to his new solution for the control of brain by mind, but instead adds a footnote on the word "interact" to explain he is using it "in a wide sense so as not to exclude psychophysical parallelism."

Indeterminism versus Self-Determination

Another main theme of Popper's philosophy, indeterminism, is applied to the mind-brain relation. In this we are in fundamental disagreement. I favor determinism of an emergent, mentalist form that follows directly and logically from my concept of mind as causal (71, 81). In contrast to Popper, I hold that every time the elements of creation, whether atoms or concepts, are put together in the same way under the same conditions, that the same new properties would emerge and that the emergent process is there-fore, causal and deterministic. To this extent and in this sense it may also be said to be, in principle, predictable, though with few exceptions, it is not so in practice. Rather than viewing the mind of man as a "first cause" or "prime mover" as does Popper (54, 57), I see the brain as a tremendous generator of emergent novel phenomena that then exert supersedent control over lower-level activities. The higher-level functional entities of inner experience have their own dynamics in cerebral activity and, contrary to Popper's interpretation of my view (57), they also "interact causally with one another at their own level as entities" (75). But the creative process is not indeterminant. The laws of causation are nowhere broken or open (excepting perhaps in quantum-level indeterminacy which is here irrelevant). It is all part of a continuous hierarchic manifold, a one-world continuum.

On these terms, human decision making is not indeterminant but self-determinant. Everyone normally wants to have control over what he does and to determine his own choices in accordance with his own wishes. This is exactly the kind of control our mind-brain model provides. But this is not freedom from causal determinacy. A person may be relatively free in this view from much that goes on around him, but he is not free from his own inner self. The emphasis here is the converse of the behaviorist contention of Skinner and others that "ideas, motives, and feelings have no part in determining conduct and therefore no part in explaining it" (6, 63). Even Skinner, however, seems in recent years to have withdrawn from his former stance to a point where his present position (64) is no longer distinctive. In that great complex of external and internal determinants that control behavior, one can pick out for emphasis either the environmental factors or those of the inner self. From my standpoint, it is the latter that especially tend to distinguish man, while the former are more characteristic of animals and increasingly so as one descends the phylogenetic scale. The self-determinants in man include the stored memories of a lifetime, value systems, both innate and acquired, plus all the various mental powers of cognition, reasoning, intuition, etc.

In any case, it has become evident that Popper's philosophical arguments for mind-brain interaction have become greatly strengthened as a result of our having countered the older, pre-1964 logic of neuroscience on its own grounds. On the other side, my own concepts of mental phenomena as causal determinants in brain processing are extended and enriched particularly in the upper linguistic and epistemological levels by the insights of Popper. I should also make clear at this point that in reading Popper's work for the first time for this occasion, I was repeatedly impressed with the great extent, particularly in regard to his general positions on epistemology, to which I feel we are in strong and warm accord. The present discussion, and concern for the impact of dualist ideology, brings a disproportionate emphasis on our relative differences.

Is Conscious Experience Causal—Or Only Its Neural Correlates?

This long chronological approach may help to clarify the difference between the view of Eccles today and his position in 1964, and similarly the sudden rise during this same period in the scientific acceptance of mental entities as explanatory constructs, as well as the recent new strengths of Popper's dualist arguments. All depend in a very critical way on the appearance of a rational alternative answer by which to refute the traditional behaviorist-materialist paradigm. The new availability of a logical contradiction to our earlier reasoning that consciousness is acausal and unnecessary for a complete account of brain function means that the multiple subjectivist pressures toward humanist views, and interpretation are no longer held at bay by behaviorist theory. The logical deterrents to dualism also are correspondingly weakened. The one new concept that appears to have the needed qualifications, and that can be said to make the interaction of such different things as physical and mental states now seem plausible, whereas up to 1964 it had seemed inconceivable, is the concept which Popper and Eccles make the main thesis of their book and that I too have proposed.

No other development is visible during this period that serves to distinguish between the causal potency of mental experience per se and that of its neural correlates, providing for the former over and above the latter, in direct contradiction to behaviorist theory. The increasingly frequent references of late to the evolutionary survival value of consciousness as evidence of its causal usefulness (31) was for many decades effectively rejected on the grounds that it is the neural correlates that are causal and have survival value, not their conscious qualities. Similarly, recent advances in cognitive and humanistic psychology now expressed in terms of the causal role of mental images and other subjective phenomena, are equally interpretable today, as in the past, on behaviorist terms that recognize the causality of the neural correlates of the subjective phenomena, but not of the subjective qualities themselves.

New developments in the mind-brain identity position, the recent "consciousness' movement in clinical and humanistic psychology, and the counterculture developments of the 1960s have all been chronologically and otherwise associated, but also similarly fail to furnish any critical reasoning that would distinguish between the causal efficacy of consciousness and that of its neural correlates, or to otherwise refute, so far as science is concerned, the long dominant materialist-behaviorist paradigm. The one development that does this and presents a logical and plausible alternative, is the modified concept of mind as a causal, functional emergent.

It is the idea, in brief, that conscious phenomena as emergent functional properties of brain processing exert an active control role as causal determinants in shaping the flow patterns of cerebral excitation. Once generated from neural events, the higher order mental patterns and programs have their own subjective qualities and progress, operate and interact by their own causal laws and principles which are different from, and cannot be reduced to those of neurophysiology, as explained further below. Compared to the physiological processes, the conscious events are more molar, being determined by configurational or organizational interrelations in neuronal functions. The mental entities transcend the physiological just as the physiological transcend the molecular, the molecular, the atomic and subatomic, etc. The mental forces do not violate, disturb, or intervene in neuronal activity but they do supervene. Interaction is mutually reciprocal between the neural and mental levels in the nested brain hierarchies. Multilevel and interlevel causation is emphasized in addition to the one-level sequential causation more traditionally dealt with. This idea is very different from those of extraphysical ghostly intervention at synapses and of indeterministic influences on which Eccles and Popper had earlier relied. The question at issue is whether this form of psychophysical interaction is fundamentally monistic as I interpret it or whether it is dualistic as presented by Popper and Eccles.

In following up this question we want to first recognize that Popper and Eccles go well beyond the given formula for mind-brain interaction to promote correlative concepts and final overall positions that are genuinely dualistic. Eccles' description of the conscious self as having supernatural origins and as something that survives death of the brain, and Popper's concepts of unembodied "world three" entities existing independently of any material substrate are distinct examples. Elsewhere in their writings, many implications can be found where they discuss the loose, open, and indeterministic nature of the liaison between mind and brain that leaves no doubt that they both have something genuinely dualistic in mind. The difficulty is that these dualistic features are indistinguishably mixed in and fused with the given theory for mind-brain interaction that itself has stood up under criticism and is regarded by many of us as being definitely monistic. Throughout their volume, it is implied that their dualistic extensions and additions are both consistent with, and supported by, the emergent causal model for mind-brain interaction.

Because this model combines features from both of the earlier classical opposing philosophies of monist materialism on the one hand and dualist mentalism on the other, I presented it at the outset as a compromise view that could be labeled either way to favor either alternative (given certain qualifications and some redefinitions). It is entirely understandable that Popper and Eccles, with their prior commitments to dualism on other grounds should try to make the new compromise as consistent as possible with their earlier thinking. I similarly could have presented it, for example, as "enlightened physicalism," "neomaterialism," "emergentist, cognitivist or mentalist materialism," "nonreductive materialism," etc. In what follows I will try to outline briefly the reasons for presenting this interactionist model as neither dualistic nor materialistic. I think it combines features that separately exclude it from both the foregoing, and that it is best recognized as a fundamentally distinct alternative. From here on it may be understood that my comments will be confined strictly to my own version of the model with which I am directly familiar. As I interpret it, this concept of the mind-brain relation not only refutes the doctrines of behaviorism and materialism, mechanistic determinism and reductionism, as Popper and Eccles correctly infer, but also and with equal force, strongly discounts dualism. By explaining conscious experience in monistic terms we undermine dualism at its source and point of strongest support, leaving for dualism only abstract arguments like those of Plato and Popper, and observations like those from parapsychology. (4)

Emergent Causation

It will be helpful as we proceed to have in mind some further concrete examples of the principles of emergent (holist) control as illustrated at different levels in some simpler and more familiar physical systems. I have used the example of how a wheel rolling downhill carries its atoms and molecules through a course in time and space and to a fate determined by the overall system properties of the wheel as a whole and regardless of the inclination of the individual atoms and molecules. The atoms and molecules are caught up and overpowered by the higher properties of the whole. One can compare the rolling wheel to an ongoing brain process or a progressing train of thought in which the overall properties of the brain process, as a coherent organizational entity, determine the timing and spacing of the firing patterns within its neuronal infrastructure. The control works both ways; hence, mind-brain "interaction." The subsystem components determine collectively the properties of the whole at each level and these in turn determine the time-space course and other relational properties of the components. The organism and its component cells and organs is another familiar example. The principles are universal.

An example I come back to for classroom illustration contrasts the programming determinants in a television receiver with the electronic and other physical interactions involved in its operation. Complete knowledge of the electronic and physical theory that enables one to fully understand, build and repair the appliance, is no help to explain why Mary struck John on channel 4, or what caused the building to collapse on 2, or the laughter on 7. There is no way that these, or the political message on channel 5, can be explained in terms of the laws and concepts of electronics. They involve a different order or level of interaction. Yet these higher order, supervening, program variables do control at each instant, and determine the space-time course of the electron flow patterns to the screen and throughout the set—just as a train of thought controls the patterns of impulse firing in the brain. The shift to a new program or to a new channel can be compared to a shift in the brain to a new mental set, focus of attention, or to a new thought sequence. Popper would presumably allocate the programs of television to a separate world (worlds within worlds?). Although the allocation of such human artifacts to a distinct separate world proves helpful in some ways and interesting in its original form as a philosophic conjecture, the current promotion of the separate worlds with a capital "W" in a true dualistic sense seems fundamentally inaccurate and misleading.

The television analogy breaks down if pushed too far, of course, in that the superimposed programs of television are linearly traceable to the recording studio, whereas the brain, by contrast, is largely a self-programming, self-energizing system. It creates its own superseding mental programs with its own built-in subjective generators calling also on a lifetime of internal memories and an elaborate built-in system of value controls and homeostatic regulators. Also, the programs passing through the television monitor lack the internal interaction and competition of those of the brain, as well as the self-developing, originative properties and an internal selector of the programs to be attended.

The conscious programs of the brain may be presumed to be created in activity that lies beyond and is different from that occurring in the geniculostriate system. The difference I envisage here is not in respect to events at the neuronal level but in more systemic, organizational, relational, configurational aspects and design features of the cerebral integration. The special central system for consciousness, or the conscious self, must include a continual registration of the changing body schema (so strong it tends to persist after limb amputations) and in reference to which sensory input is consciously perceived, as well as a feeling for the volitional command of the system, and the relating of both these to sensory inputs, to memory, and to emotional values and homeostatic needs. The conscious attentional component in this central metasystem may be only a small surface feature of the whole vast complex of cerebral integration. The crucial features of the central self system are presumably innate in each species and largely preorganized independently of sensory input.

It is important to recognize that the term interaction applies in these examples only in the general sense in which it has been used in the history of psychology and philosophy to imply a causal influence between mind and brain. I have stressed that the term interaction is not to imply that the mental forces intervene in, or disturb or disrupt the physiology or chemistry of the brain, but only that they supervene, like TV programs over the electronic processes. No interruption or violation of the laws of physiology is involved. I infer that Popper and Eccles also use the term mostly in the same way and only rarely in the more specific sense of an actual disturbance of physiological events, as some seem to have misinterpreted their meaning.

Unembodied Minds?

Given our original description of the theory and its consistent reiterations, along with illustrative examples like the foregoing, it is not easy to understand how this concept of the mind-brain relation could be taken as support for dualism. First, it fails to satisfy the classic philosophic definition of dualism as two different forms or states of existence neither of which is reducible to terms of the other. Our theory describes the mental states as being built of, composed and constituted of physiological and physicochemical elements, and thus, in the sense of the definition, reducible to these. It needs to be explained here that much confusion has arisen from the use of this term "reducible" in two quite different senses in different contexts. In common usage a building is said to be reduced to rubble by an earthquake. This is denied, however, in philosophic, holist-reductionist dispute on the contention that in the reduction process, even with careful disassembly, the building as such has been lost and therefore has not been, and cannot be, in principle, reduced to its parts. It is only in this latter specialized sense, and not in the common sense of the above definition or dictionary usage that I describe the mental events as not reducible to brain physiology.

The reason that mental or other entities cannot be thus reduced to their parts may be understood more easily if one thinks of a given entity not as a system of just material components, but as a combined space-time-mass-energy manifold. Think of space being bent around and molded by the material parts and time as similarly defined by events in temporal and moving systems with the space-time components both arranged also in vertical nested hierarchies corresponding to and filling in around the material elements and defined by their relative positions and timing. The process of reducing an entity to its material components, physically or conceptually, inevitably destroys the space-time components at the affected level. These last components from the space-time manifold, interfusing with, shaped by, and demarcated by the material components, are highly critical in determining the causal and other distinguishing properties of any sysem as a whole. The spacing and timing of the parts with reference to one another largely determine the qualities and causal relations of the whole, but the laws for the material components fail to include these space-time factors. Attempts to recognize them in so-called collective and cooperative effects tend to fall short of an adequate recognition of the basic importance of the space-time elements. This is why quantum mechanics is of little help in explaining physical reality at orders much above the quantum level.

None of this is to reject the value of reduction as a method in science or as a means to gain understanding in general. The properties of any entity are determined largely (but not entirely, and in some cases more than others) by the properties of its parts. It obviously helps enormously, as a rule, to know how and of what anything is composed. Further reduction to the composition of the parts of the parts, and so on, becomes increasingly less explanatory of operations at the higher, starting level. Though brain quarks and gluons are not of particular relevance to behavioral science, one can expect that in many respects brain physiology in its upper dimensions may become to behavior and cognitive processing what molecular theory is to chemistry. It is only the reductionist reasoning that therefore things can be reduced to "nothing but" their parts that is rejected, or that all science can be reduced, in theory, to a basic unity in one fundamental discipline, or that the "essence" of anything is to be sought in its components.

Along with the failure to qualify as dualism by definition, our proposed mind-brain model also is nondualistic in that it makes mind and brain inseparable parts of the same continuous hierarchy, the great bulk of which, by common agreement, is not dualistic. It becomes illogical to make a special exception of the principle at the one level of mind and not at those above and below. In the proposed scheme, one can proceed continuously in the same universe of discourse, following the path of evolution, from subatomic elements in the brain up through molecules, cells and nerve circuits to brain processes lacking or having conscious properties and on upward through higher compounds all within the one this-world mode of existence.

Dualism would seem to be further contradicted by our description of subjective meaning as a functional derivative rather than as a brain copy or a spatiotemporal transform. As an emergent functional attribute of brain activity, conscious experience is inextricably linked to, and inseparable from, the functioning brain. It is only in the functional relations within the matrix of brain processing that the subjective qualities appear and have their meaning. The subjective effects are generated by, and exist only by virture of, brain activity. Even where higher order mental forms are compounded of lower level mental entities, as we assume to be the case, the entire hierarchy is still embodied in, dependent on, and inseparable from the physiological substructure.

Much the same solution to the mind-brain problem has been arrived at recently by MacKay (43) who presents it in the more restricted terminology of information theory using for illustration the example of goal-directed operations in a computer. The same example was offered by MacKay in 1964 couched in "dual aspect" theory when he held the view (most favored in neuroscience at that time) that the mental and the physical are complementary aspects of one and the same process where "no physical action waits on anything but another physical action" (MacKay 1966:438). In those years, MacKay granted the physical determinancy for the central nervous system holding conscious brain activity to be predictable in principle, in objective terms from a knowledge of the precedent physical determinants (provided one did not reveal the prediction of a sequence to a person involved in the prediction), The emergent nature of the mental controls as we now conceive them in a vertical or nested hierarchy, and the manner in which they supersede, rather than merely parallel as an inner aspect, the physiological determinants were missed by MacKay in 1964 but apparently are accepted in his 1978 version, along with a new recognition of the causal efficacy of consciousness. These changes seem now to bring our respective views into rather close accord with regard to those features most directly relevant to the mind-brain problem.*

Apparently unacquainted with the history of these conceptual developments and the original concepts from which Popper and Eccles argue, Mackay (43) misinterprets the kind of interactionism they have in mind and then finds it "astonishing" how close and natural a fit can otherwise be made between theirs and his own description. Preserving consistency with his earlier position, MacKay is inclined to emphasize, more than I, the extent to which the proposed alternative is a physicalist rather than a mentalist view. It has to be remembered in this regard that whereas the programs of the computer or television analogies are conceived in phsyical terms, those of the brain have always been described as *mental* with subjective properties defined as a contrast to the physical or material. In any case, I fully agree, however, that the arguments and evidence advanced in support of dualist-interactionism in the Popper and Eccles volume are very much open to the kind of alternative interpretation that we propose.

The New Mentalism and Materialist Philosophy

The explanation of mind in the foregoing terms as an organizational functional property of brain processing, constituted of neuronal and physicochemical activity, and embodied in, and inseparable from the active brain has led to an impression in some cases that this should properly be interpreted as therefore an essentially materialist view. Some further reasons for defining it instead as mentalist (or cognitivist) can be outlined as follows: The

* Although MacKay's 1978 article was subject to the interpretation here described, this apparently was not MacKay's intention. In his later 1980 book, *Brains, Machines, and Persons*, MacKay makes it clear that he has not deviated in his thinking from his earlier 1966 dual aspect position. In the 1980 account he continues to regard mind and brain as complementary aspects of one and the same thing which are likened respectively to an "inside" and an "outside" story that run in parallel and are correlated but do not interact. His parallelistic "two languages," "two logics" position preserves a strictly physicalist determinism in brain function without allowance for the kind of mentalist determinism of neural activity conceived here.

principal feature of this model is the new recognition it gives to the primacy of subjective mental phenomena in scientific explanation and the higher level control role accorded mental or cognitive phenomena as causal determinants, over and above their neural correlates. I characterize it as placing "mind back over matter," and as "a scheme that idealizes ideas and ideals over physical and chemical interactions, nerve impulse traffic, and DNA. It is a brain model in which conscious mental psychic forces are recognized to be the crowning achievement of some five hundred million years or more of evolution" (73). As such, it conforms to the common textbook and lay definitions of the terms mental and mentalism. The subjective qualities are recognized to be real and causal in their own right, as subjectively experienced, and to be of very different quality from the neural, molecular, and other material components of which they are built. Because mind and matter, the mental and the physical, have long been defined as direct contrasts and given meaning in terms of their opposites, this proposed recognition of the causal primacy of subjective mental qualities would seem to logically exclude materialism.

In particular, the present position represents a direct refutation of what materialism had long come to stand for over many decades in science, philosophy, and humanist thinking generally. Materialist behaviorism asserting the principle that ideas, motives and feelings have no part in determining conduct and therefore no part in explaining it (6) had gone, in the extreme, to denying even the existence of consciousness in any form and, at the least, to denying as a founding central premise any causal efficacy of conscious or mental forces in brain processing. Materialist philosophy and the so-called psychophysical identity theory was being advanced during the 1960s on the contention that "man is nothing but a material object, having none but physical properties," and "science can give a complete account of man in purely physico-chemical terms" (1). The "Unity of Science" movement, closely aligned with identity theory, held that the laws of science can all be reduced eventually, in principle, to the laws of a single basic discipline (12, 26, 48). Physical science was seeking answers to all nature in terms of "the four fundamental forces" with hopes for a further unifying field theory to describe the essence of reality. My view arose in the mid-sixties, in direct opposition to all of these related materialistic, mechanistic, and reductionist trends.

In the meantime, mind-brain identity theory, which has become the strongest thrust in materialist philosophy, has undergone substantial changes during the last decade. In its initial form as a semantic twist to the old "double aspect" view that goes back at least to Spinoza, it was described as a "dual access" or "double language" theory (27) and was strongly reductivist. In particular, it held that a complete account of brain processing is possible, in principle, in neural terms only without resorting to subjective language or mental terms. Unlike the epiphenomenal view, or the emergent, double aspect, or interactionist views, identity theory itself seems to provide no new concrete concepts to the mind-brain problem, only different semantic approaches. My introduction in the mid-sixties of the opposing view of consciousness as a nonreductive emergent with causal potency and downward control has been followed by a spate of semantic transformations in identity theory in which a new emphasis is put on the causality of consciousness and on emergent concepts under terms such as organizational, configurational, holistic, collective, and the like (30, 49, 65, 89, 90).

In all cases the changes appear to bring these two initially contrasting approaches into closer convergence. The argument from identity philosophy today seems accordingly to be not as much that my emergent determinist view is incorrect, but rather that this is what identity theory actually should have been taken to imply all along. We thus have the curious result that my compromise mind-brain model is today being identified with materialism on the one hand, and with dualism on the other.

Finally, in defense of the mentalist rather than the materialist designation, I would only add the following: if there is anything in this world that has been commonly defined as a contrast to the material or physical, it is the intangibles of conscious experience. The psychological contents of mind from their first recognition in language, philosophy, and science, have been treated by tradition as opposites of physical and material in the mind-matter dichotomy. Accordingly, a position can hardly be called materialist if its very essence and reason for being is a new antimaterialist stress on the existence and functional primacy of mental phenomena and their role as high-level causal determinants in brain function, obeying laws that are different in kind from those of their constituent material, neuronal, and electrochemical processes. A mentalist is defined in behavioral science as one who, in opposition to behaviorist doctrine contends that mental entities and laws are involved in determining behavior and are needed to explain it. The concept of consciousness as a causal emergent has been presented from the outset as a view that restores to science the common sense impression (overruled during the behaviorist-materialist era) that we do indeed have a mind and mental faculties over and above, and different from our brain physiology—just as we have cellular properies that are over and above and different from their molecular constituents.

The distinction between the mentalist philosophy and that of materialism or behaviorism, though important within psychology, is less critical overall than that between monism and dualism. If common usage in the long run should tend to favor the stretching of the meaning of materialism and/or physicalism to encompass mental phenomena in the causal, emergent, embodied, nonreductive form we now envisage, there would be no great loss provided there was no resultant confusion in regard to the actual conceptual changes themselves and their new implications and connotations. Of all the questions one can ask about conscious experience, there is none for which the answer has more profound and far-ranging implications than the question of whether or not consciousness is causal. The alternative answers lead to basically different paradigms for science, philosophy, and culture in general.

If the concern with terminology begins to seem overdone, it should be remembered that labels and their connotations and the right hemisphere impressions they carry are often more important in human decision-making than are the more precisely formulated logical concepts and facts they stand for. When Popper and Eccles, representing modern philosophy and neuroscience, jointly proclaim arguments and beliefs in dualism, the supernatural, and unembodied worlds of existence, the repercussions quickly extend beyond professional borders to influence attitudes and faith-belief systems in society at large. The result has been a major setback for those of us who see hope for the future, and for the very aims and ideals that I think Popper and Eccles strive for, to lie in replacing old dualist perspectives, values and beliefs, dualist theologies and related mythological, supernatural guidelines of the past with a new unifying holistic-monistic interpretation of reality as an ultimate reference frame for transcendant value and higher meaning.

7

Changing Priorities

Toward a Union of Science with Ethics and Religion

Bringing various threads together in this final chapter, we return to the starting argument that rising disaster trends around the world are traceable primarily to misguided human value priorities, and that the most effective prescribed remedy is to bring our value-belief systems more into tune with this-world reality. The relevance, rationale, and various convergent lines of reasoning, behind this conclusion are here presented in the context of some common practical perpectives we encounter today in neuroscience and in society at large.

Shaken Confidence in Science

We used to say that there are two kinds of scientists: those fired-up by a problem and searching for methods to get answers, and those highly trained in some method who are searching for some amenable problems. While most of us line up somewhere between these extremes, there is much to be said for, at least in principle, giving preference where possible to problem priorities over methodology. What follows is, above all, problem-oriented and attuned throughout to the query: "What difference will our science make—especially ten, twenty, or more years from now?"

In terms of government funding and in other respects, it has become apparent that the overall federal rating for neuroscience is not so high today as it was prior to the federal budget reforms of the early seventies. Nor does it appear that the decline is something temporary from which funding can be expected soon to recover. Nor is it restricted to neuroscience. Science in general has been affected with certain exceptions such as cancer-related, energy-related, and other select projects where a major application to current quality-of-life problems is obvious. These changes in relative funding can be assumed to reflect real changes in social priorities and in society's collective judgment of the importance of science and what it contributes. We read in *Science* (62) of the "public disillusionment" and "today's more jaundiced view" of science and that "faith in the beneficence of scientific endeavor and the promise of technology has been steadily eroding."

An underlying cause for these changes can be seen in the new and growing recognition of mounting world crisis problems which science is accused of having helped to create and which in addition are complicated by social value problems for which science is apparently unable to provide answers. When the quality and even survival of civilized society is potentially in threat, what difference does it make to Congress or the public whether we find some new nerve connections in the brain, some new transmitters or receptors, and so on? Even the ever strong humanitarian appeal of medical advancements that might eventually save hundreds of thousands of lives does not fully escape the new unspoken perspective of a world already afflicted with population imbalances in the hundreds of millions. The overwhelming priority of the growing demands of contemporary social problems was already perceived in the late sixties (50) to be of sufficient magnitude and urgency to warrant mustering the scientific community to an all-out crash attack with the warning that to continue the practice of "science as usual" was morally indefensible.

Although little has happened in the interim to reduce the specter of global breakdowns, it seems that a lot has happened to discourage public feeling that science and improved technology can be counted on to bring solutions. While the world's production per capita of most major products of the basic biological source systems has already peaked and started on a long, downward trend (8), world population continues to rise at the rate of six million people per month with predictions of inevitable social turmoil as peoples and nations grow more desperate.

Improved Technology Is Not Enough

Earlier hopes that science might rise to the occasion with "green revolutions" and other technological answers begin to fade. Science and improved technology, we come to realize, merely make it possible to better maintain more people in better style, for a while, until new limits are reached and the same problems reappear, along with new ones, and all on a greatly magnified scale. Science and advanced technology, whether medical, agricultural, military, energy-related, etc., are now seen, in the long run, to put us in an escalating vicious spiral of runaway technology-population-energy-pollution increments, in which we are now firmly entrapped. This is no reflection on science or technology per se, of course. As we say, utopia is tomorrow's level of technology combined with the population levels of the nineteenth century. Remedial suggestions, however, that might in any way involve the highly sensitive, value-laden factor of population controls promptly raise a host of moral issues and value conflicts for which, again, science, it is held, does not provide answers.

Futurists and common sense concur that a substantial change, world-wide, in life style and moral guidelines will soon become an absolute necessity. On a planet of finite resources, the laws and mores of a freely increasing population must eventually be replaced by those of a regulated population, and the sooner this inevitable shift occurs, the better for the residual quality of the biosphere. In short, it becomes increasingly evident that the prime, urgent need of our times is not for more science and improved technology, medical, agricultural, or otherwise, but for some new ethical policies and moral guidelines to live and govern by that will work against overpopulation, pollution, depletion of resources, and so on.

Once this conclusion is perceived, the common tendency is to discount science and look elsewhere for answers. Problems that resolve to basic issues in ethics and morality are traditionally supposed to be beyond science. It is hardly surprising, therefore, that public faith in the promise of science and technology has been steadily eroding.

A Different Approach

In what follows I try to defend a position directly counter to the above, which would, in effect, not only restore to science any loss in public favor, but would go further to give science, and the scientific endeavor generally, a change of image and a higher societal role. On the proposed terms, science becomes the prime hope for escape from the vicious spirals of advancing civilization, but not for reasons usually cited. A different approach to the public support and role of science is envisioned in which science is upheld, not because it begets improved technology, but because of its unmatched potential to reveal the kind of truth on which faith, belief, and ethical principles are best founded. In the world view and truths of science we will find the best key to valid moral guidelines. The arguments are adapted to today's problems and grow stronger, not weaker, as current global conditions worsen. Even basic "pure" research and the practice of "science as usual" emerge on the proposed terms with a heightened social and moral approval.

The usual appeal to medical, educational, technological, and other direct benefits is by-passed and our bets are placed instead on certain less obvious human value implications that stem from the sciences. Particularly relevant are recent changes in concepts relating to the mind of man, the nature of the conscious self, freedom of choice, causal determinacy, and to the fundamental relation of mind to matter and to brain mechanism. Some of man's most profound concerns are involved, i.e., whether consciousness is mortal or immortal, cosmic or brain-bound, or reincarnate, and the like. It is in terms of the humanistic implications along these and related lines that mind-brain science has always had its special interest and greatest meaning. Ideologies, philosophies, religious doctrines, world models, value systems, and more will stand or fall depending on the kinds of answers that brain research eventually reveals. It all comes together in the brain.

In brief, recent conceptual developments in the mind-brain sciences are seen to bring changes in world view perspectives that revise the ultimate criteria and frame of reference for determining human value priorities and resolving value differences. A broad shift of conceptual framework regarding science and human values is involved. Promising prospects can be seen, especially as these changed perspectives apply in those global crisis areas wherein lie today's most serious threats to the quality of life, and where different outcomes in the resolution of value conflicts tend to have tremendous social consequences. For example, even a slight shift in the delicate balance of value perspectives on the abortion versus right-to-life issue could mean the difference, in itself, of literally many millions of lives within the next few years, with enormous secondary impacts as well—and all compounded on future generations. Similar wide-ranging, quality-of-life consequences result from a shift of values in regard to energy-environmentalist issues, species' rights, and other worldwide concerns.

It is our present contention that a scientific approach to both the theory and the prescription of ethical values is not only feasible, but is by far the best way to go, offering the most promising, perhaps only, visible hope for future generations. The supporting arguments have already been expounded in some detail elsewhere and may be found in the preceding chapters and their references at the end of this volume. Rather than assume prior knowledge or laboriously restate the reasoning in different words, it is more expedient for present purposes to simply relist here some of the principal postulates, propositions, observations, and inferences as excerpted with minor changes from the previous chapters and other accounts. Because the subject matter ranges somewhat afield from conventional disciplines, and becomes, again, of the great urgency and magnitude of the ultimate concerns at stake, overlap and redundancy are risked rather than the reverse. Some attempt at logical ordering is maintained, but the cross logistics mount rapidly, and a quick grasp of the whole may be found preferable to a logical sequential approach.

Collective Postulates, Inferences, and Propositions:
A Recapitulation

SUBJECTIVE VALUES IN OBJECTIVE PERSPECTIVE

• In addition to their commonly recognized significance from a personal, religious, or philosophic standpoint, human values can also be viewed objectively in causal control terms as universal determinants in all human decision making. All decisions boil down to a choice among alternatives of what is most valued, and are determined by the particular value system that prevails.

• Human values, viewed in objective, scientific perspective, stand out as the most strategically powerful causal control force now shaping world events. More than any other causal system with which science now concerns itself, it is variables in human value systems that will determine the future.

• Any given brain will respond differently to the same input and will tend to process the same information into quite diverse behavioral channels depending on its particular system of value priorities. In short, what an individual or a society values, determines very largely what it does.

• As a social problem, human values can be rated above the more tangible concerns such as those of poverty, pollution, energy, and overpopulation on the grounds that these more concrete problems are all manmade and are very largely products of human values. Further, they are not correctable on any long-term basis without effecting adaptive changes in the underlying human values involved.

• The human value factor in biospheric controls stands out as the primary underlying basic cause of most of today's difficulties. The more strategic way to remedy mounting adverse world conditions is to go after the social value priorities directly in advance, rather than waiting for man's values to change in response to changing external conditions. Otherwise we are doomed to live always on the margins of intolerability, for it is not until things begin to get intolerable that the voting majority gets around to changing its established values.

• Recent developments in the mind-brain sciences eliminate the traditional dichotomy between science and values and support a revised philosophy in which modern (nonreductive) science becomes the most effective and reliable means available for determining valid criteria for moral value and meaning.

VALUE THEORY

• A science of values in the context of decision theory becomes conceivable extending into all branches of behavioral science and forming a skeletal core for the social sciences.

• The seemingly endless complexity of human values is greatly simplified by viewing values in hierarchic structures which are goal-dependent, and further, by restricting attention to those areas that are involved in social conflict.

• The innate components of the value structure, which include inherent psychological as well as biological values, can be treated largely as a common invariant denominator of human nature, allowing the focus of attention to be directed to problems of the acquired, cognitive, ideological values where the major sources of value conflict arise.

• On analysis, values are found to be correlates of directed activity. They are always relative to some purpose, goal, or aim, explicit or implicit, and structured in goal-dependent hierarchies. Any concept or belief regarding the purpose and value of life as a whole, once accepted, then logically supersedes and conditions the entire hierarchy of value priorities at subsidiary levels. Values at the ideological plane become ordered and ethical issues judged in accordance with the conceived ultimate purpose of life as a whole. This latter will logically imply at the same time an associated world view or universe scheme that is consistent.

• Because of the hierarchic structure of values, the search for improved ethical guidelines can be narrowed to the search for what ought to be most valued. This in turn leads to problems of the highest determinants of value priorities—the life-goal, world model concepts and beliefs that lie at the heart of the problem of moral judgment and logically condition the value structure at all levels.

• Societal values, especially of the kind people disagree on, are always dependent upon, and relative to, some general frame of reference containing the premises, beliefs, and presuppositions on which the reasoning about priorities rests. The question may be raised: What makes one reference frame superior or supersedent to another? and then: Is there some ultimate frame of reference for values that could logically and rightly be accepted and respected by all countries, cultures, governments, and creeds, and by mankind in general, as the final supreme standard when it comes to judging ethical priorities, resolving value conflicts, and as a guideline for human judgment generally and international decision making in particular? The practical need for some such unifying global standard becomes more and more evident for things such as world population control, conserving world resources, protecting the oceans and atmosphere, and for various other modern world problems that increasingly require united effort on a global scale.

• What is needed ideally to make decisions involving value judgments is a consensus on some supreme comprehension and interpretation of the universe and the place and role within it of man and the life experience.

DEPENDENCE ON MIND-BRAIN CONCEPTS

• Beliefs concerning the ultimate purpose and meaning of life and the accompanying world view perspectives that mold beliefs of right and wrong are critically dependent, directly or by implication, on concepts regarding the conscious self and the mind-brain relation and the kinds of life-goals and cosmic views which these allow. Directly and indirectly social values depend, for example, on whether consciousness is believed to be mortal, immortal, reincarnate, or cosmic, and whether consciousness is conceived to be localized and brain-bound or essentially universal.

• Recent developments in mind-brain theory revise the ultimate criteria and our ultimate frame of reference for determining value priorities. Problems of values, ethics, and morality (questions of what is good, right, and ethically true and of what ought to be) become, in these revised terms, something to which science, in the most profound sense, can contribute fundamentally and in which science should be actively and responsibly involved.

• Current concepts of the mind-brain relation involve a direct break with the long-established materialist and behaviorist doctrine that has dominated neuroscience for many decades. Instead of renouncing or ignoring consciousness, the new interpretation gives full recognition to the primacy of inner conscious awareness as a causal reality.

• The phenomena of conscious experience are conceived to play an active, directive role in shaping the flow patterns of cerebral excitation. Instead of being parallelistic and acausal, consciousness in the present scheme becomes an integral part of the brain process itself and an essential and potent constituent of the action. Consciousness is put to work and given a use and a reason for having been evolved in a physical system. Subjective phenomena including values are brought into the causal sequence in human decision making and behavior generally and thus back into the realm of experimental science from which they had long been excluded.

• The seemingly irreconcilable dichotomies and paradoxes that formerly prevailed with respect to mind versus matter, determinism versus free will, and objective fact versus subjective value become reconciled today in a single comprehensive and unifying view of mind, brain, and man in nature.

• The swing in psychology and neuroscience away from materialism, reductionism, and mechanistic determinism toward a new, monist, mentalist paradigm restores to the scientific image of human nature the dignity, freedom, responsibility, and other humanistic attributes of which it has long been deprived in the materialist-behaviorist approach.

• A nonreductive holistic world model and interpretation of physical reality is supported in which the qualitative pattern properties of all entities are conceived to be just as real and causally potent as those of their components. This preservation of the qualitative value and pluralistic richness of physical reality stands counter to the common tendency to correlate science with reductionism.

TOWARD MERGENCE OF SCIENCE AND VALUES

- Instead of separating science from values, the current interpretation leads to a stand in which science—in its purest sense as a means of revealing an understanding of man and the natural order—becomes the best source, method, and authority for determining the ultimate criteria of moral values and those ultimate ethical axioms and guidelines to live and govern by.

- The classic fact-value and naturalistic fallacies of philosophy logically dissolve in the context of cerebral processing. The operations of the brain are already by nature richly replete with established values and value determinants, both inherent and acquired, with the result that incoming facts regularly interact with and shape values. The resultant value system, including conceptions of what ought to be, is determined in very large part by the factual input.

- Changing to an ethic based on science would entail in large part a substitution of the natural cosmos of science for the different mythological, intuitive, mystical, or otherworldly frames of reference by which man has variously tried to live and find meaning.

- The aim is not to eliminate value controversy and differences of faith and opinion but only to bring these into a domain set by an agreed-upon frame of reference supported by science—not with the idea that scientific truth is absolute or beyond question but only with a conviction that it does represent the best and most reliable, credible, and dependable approach to truth available.

- Once science modifies its traditional materialist-behaviorist stance and begins to accept in theory and to encompass in principle within its causal domain the whole world of inner, conscious, subjective experience (the world of the humanities), then the very nature of science itself is changed. The change is not in the basic methodology or procedures, of course, but in the scope and content of science and in its limitations, in its relation to the humanities and in its role as a cultural, intellectual, and moral force. The kinds of interpretations that science supports, the world picture and attendant value perspectives and priorities, and the concepts of physical reality that derive from science all undergo substantial revisions on these new terms. The change is away from the mechanistic, deterministic, and reductionistic doctrines of pre-1965 science to the more humanistic interpretations of the 1970s. Our current views are more mentalistic, holistic, and subjectivist. They give more freedom in that they reduce the restrictions of mechanistic determinism, and they are more rich in value and meaning.

• Accepting as the ultimate value of what man generally has held most sacred, namely, the cosmic forces that made, move, and control the universe and created man, and interpreting these in conformance with the worldview of science, one emerges with a value system that includes a strong reverence for nature, promoting the values of the recycle philosophy, population regulation, and protecting the environment and for progressive enhancement in the quality of existence generally.

• In the eyes of science, to put it simply, man's creator becomes the vast interwoven fabric of all evolving nature, a tremendously complex concept that includes all the immutable and emergent forces of cosmic causation that control everything from high-energy subnuclear particles to galaxies, not forgetting the causal properties that govern brain function and behavior at individual, interpersonal, and social levels. For all of these, science has gradually become our accepted authority, offering a cosmic scheme and view of the human psyche that renders most others simplistic by comparison and which grows and evolves as science advances.

• Science becomes man's best channel for gaining an intimate understanding of and rapport with those forces that made, move and control the universe and created man. None of this is to suggest that science take on the functions of religion, but only that there might be mutual and other benefits from a fusion of the two.

• The future of science will be very different depending on whether science is recognized in the public mind to have competence in the realm of values. Reciprocally, and of far greater importance, the future of society itself will be very different depending on whether its value perspectives are shaped by the truths and world view of science or by other alternatives that now prevail.

Key to Quality Survival

Implicit in the foregoing is the conclusion that our top social priority today is to effect a change worldwide in man's sense of value. This translates by hierarchic value theory into a change in what is held most sacred. What is needed, more specifically, is a new ethic, ideology, or theology that will make it sacrilegious to deplete natural resources, to pollute the environment, to overpopulate, to erase or degrade other species, or to otherwise destroy, demean, or defile the evolving quality of the biosphere. This is exactly what is found to emerge from our current approach to the theory and prescription of human values. Relying on the kind of truth supported by science we arrive at an ethic that promotes an ultimate respect for nature and its creative principles, including those of its peak thrust into the highest esthetic, emotional, intellectual, and spiritual reaches of man's mind, along with corollary value criteria which, if applied worldwide, would promptly set in motion the kinds of corrective legislation and other trends and pressures that are needed to remedy looming global disaster conditions.

On the terms proposed, the utility of science takes a different form. Society would look to science not only for new technology and objective knowledge, but more importantly, for the criteria of ultimate value and meaning. Each advance of science brings increased comprehension and appreciation of the nature, meaning, and wonder of the creative forces that move the cosmos and produced man. Even "science as usual" gains, in this context, a heightened social significance and moral support. The special key role of neuroscience and brain research will be readily apparent.

It remains to further stress a point already implied, namely, that for science to fully qualify and function effectively in this changed role, it will be necessary that we reform an entire mode of scientific thinking long referred to under the general rubric of "scientific materialism." Moves to abide by the truths of science, as opposed to unproven claims from other sources, have had sporadic support since science began. What is new today is the shift in science from reductive physicalism to a holist-mentalist paradigm and the changed interpretations and perspectives that this brings. Among traditional views that consequently require correction is that predicating the impotence of science in regard to value judgments along with much of the doctrine associated with reductive mechanistic determinism that for many decades has characterized science and our scientific outlook. This was the thinking of Karl Marx and is the reason that the more materialistic and animalistic aspects of human nature are put first in Soviet philosophy, before man's more idealistic, spiritual components. The issues at stake are not minor. They involve (in addition to the humanist implications stressed above) not only the public image of science, the relation of science to human values, and the kinds of values science upholds, but also some of our more basic concepts within science itself concerning physical reality, mind and matter, and the nature of causation.

Different Forms of Causation

The various issues converge on opposed views of casual determinism that are basic and central to everything stated thus far. One view holds that the causal forces and laws operating in nature are fully explainable in purely physical terms and are, in principle, ultimately accountable on the basis of quantum theory, the elemental forces of physics, or in some more unifying field theory eventually to be found. Physicalist, i.e., materialist determinism is assumed to prevail throughout nature in this view, and all higher level interactions, including those of the brain, are presumed to be reducible and accountable in terms of the ultimate fundamental forces of physics.

Opposed to this long dominant physicalist-behaviorist interpretation is the view I uphold here and which has recently been gaining increased acceptance particularly in the behavioral sciences. It contends that the higher forces and laws of causation, as seen for example in classical mechanics, in physiology, and in brain function and behavior, cannot be fully explained by the laws of quantum mechanics or by the mechanics or laws of any other ultimate physical force or field. The higher entities and their causal properties and laws of interaction are conceived to be causal realities in their own right and not determined completely (though they are in part) by the causal laws and properties of their components. The larger, higher, more molar properties are perceived to be just as real, just as causal, and in many ways to be of more importance than are the more basic physical properties of their subsidiary components. In this view the fundamental forces of physics are only building blocks used in creating bigger, more competent entities and forces. The patterning of the building components, i.e., their arrangement in space and time, becomes a distinctive key factor in making things what they are, and is not determined solely by the properties of the parts themselves.

To attempt to explain an entity in terms of its parts and then the parts in terms of their parts and so on, results in an infinite regress in which one is left at the end trying to explain everything in terms of next-to-nothing. At each step of the way critical pattern components of causality are lost and the explanation becomes less and less complete at each lower level. To attempt to include, even in principle, the pattern factors, i.e., the space-time components, by invoking the "interactions of the parts," or the "organizational relations," at each step, sounds good but is empty lip service. We have no science for the space-time components, no science for the collective form in which these are present at each level of the infrastructure. Even the relatively ultra-simple interactions involved in the classic "three-bodies problem" are formidable.

Our present view holds further that when a new entity is created, the new properties of the entity, or system as a whole, thereafter overpower the causal forces of the component entities at all the successively lower levels in the multinested hierarchies of the new infrastructure. In other words, whenever an entity joins forces with others to form a new whole, the position that it is forced to take in the universe and its subsequent course through time and space and its eventual fate are thereafter determined more conspicuously by the new properties of the system as a whole than by its own original properties. A degree of self-determinacy is lost to the parts as soon as the higher powers of the new whole become superimposed. Although the causal forces at the lower quantal, atomic, molecular levels in the infrastructure continue to operate in full force as usual, they are enveloped, encompassed, overwhelmed, superseded, supervened, and outclassed by the new causal properties that emerge in the whole. Evolution, in the course of compounding new compounds continuously adds new entities and new phenomena that embody new qualities, new causal forces, and principles with new scientific laws and control properties.

The new emergent phenomena, not reducible to their parts and deserving to be recognized as causal realities in their own right, are in many respects more powerful and dominant features of reality than are the lower properties of the components. Instead of a universe completely controlled by quantum mechanics and the basic forces of physics, science presents, by this interpretation, a universe controlled by a rich profusion of qualitatively diverse emergent powers that become increasingly complex and competent. Any randomness, chance, caprice, or chaos that may be operating at the quantum level, as modern physics insists, gains little expression because it is effectively superseded and controlled by higher level forces that are anything but random. The higher levels involve much more than mass probabilities. The creative, interlocking web of evolving nature is not blind or chancelike but becomes, as it progresses, rich in irreversible, directional, ever more complex constraints that tend to keep things moving in a trend toward higher and more competent forms.

In the brain, controls at the physicochemical and physiological levels are superseded by new forms of causal control that emerge at the level of conscious mental processing, where causal properties include the contents of subjective experience. Causal control is thus shifted in brain dynamics from levels of pure physical, physiological, or material determinacy to levels of mental, cognitive, conscious, or subjective determinacy. The flow of nerve impulse traffic and related physiological events in a conscious process is no longer regulated solely by events in kind but becomes caught up in, enveloped, and moved by the higher mental controls, somewhat as the flow of electrons in a television set is moved and differentially patterned by the program content on different channels. Just as the programming variables of a TV monitor have to be included in order to account for the electron flow pattern of the system, so also in the brain the subjective, mental variables of cerebral function have to be included to give a full account of the flow patterns of neural excitation. The mental events of conscious experience and the physicochemical events of the infrastructure are not conceived to be in parallelistic relation like that of "two languages," "two logics," or of "two complementary aspects of one and the same situation" in which a "purely physical determinacy" of the central nervous system is preserved as dual aspect theory (44) would have it. This shift from a causal determinacy that is purely physical to one that includes conscious subjective forces that supersede the physical—in other words the shift from a materialist, reductionist, mechanist paradigm to a holistic, mentalist paradigm—makes all the difference when it comes to using the "truths" of science in building ethical values.

Marxism Inverted

In trying to assess possible social repercussions and the outcome of a societal shift to an ethic founded in science, it would be unfortunate and misleading if one were to rely on Marxism and the Communist world as an example. According to our latest mind-brain theory and its implications, Marxist-Communist doctrine is founded on some basic, dated errors in the interpretation of science and of what science stands for and implies in reference to human nature and to social and worldview perspectives. As a result, the kinds of values upheld in Marxist doctrine are in many respects almost the diametric opposite from those which emerge from a scientific approach on our present-day terms (3, 61).

If the growing competition between Communist and free world countries is to continue to be in part a battle for men's minds—a battle of ideologies and beliefs, and of conflicting values—it may be worth a few words in closing to point out some of the ideological value differences that result even though the intent in both cases is to exclude dualist otherworldly answers in favor of the this-world truths of science. Some of the starting differences in basic philosophy include the following:

1. First and foremost, Communist doctrine developed within the long accepted—but now largely overthrown—view

that science, of necessity, leads to and supports a materialist philosophy that rejects subjective mental phenomena as causal and predicts instead a purely materialist determinancy for brain and behavior.

2.   The doctrine of materialist reductionism was predominantly accepted in application to nature in general and to human behavior in particular. (But see footnote.*)

3.   In relation to the foregoing, Marxist philosophy failed to recognize the key principle of downward causation, i.e., the causal control that higher emergent properties in any entity, whether a society or a molecule, invariably impose over the lower properties of their infrastructure.

4.   Marxism developed in the absence also of any theory that serves to resolve the is-ought fallacy or the traditional dichotomy that has heretofore kept science and human values separate.

5.   Also lacking in Marx's time were the free will concepts we have today that liberate individual and social decision-making from mechanistic determinism.

6.   Marx opted for a strongly homocentric value system that makes man the measure of all things and gives precedence to the basic material needs of man over the quality of the biosphere, as well as over man's higher psychological needs. There is no justification in science for this choice and it is, in some respects, an inversion of nature that puts the welfare of a part of a system above that of the system as a whole.

Basically, according to Marx, what counts in human affairs and changes the world are man's actions in fulfilling his material needs for subsistence, not man's idealism, philosophy, or ideology. He emphasized that the materialistic-animalistic needs must come first and that the higher human pursuits are built upon and depend on the more basic components. On the other hand, Marx failed to appreciate that the higher idealistic properties in man and society, once evolved, can then reciprocally supersede, encompass,

---

* According to recent reports the Soviets are currently engaged in a massive campaign to revamp and update their official philosophy in directions advocated here, without, of course, admitting to the people any major past error and searching out the historical bases and rationale beyond Marx in other Soviet writings. When it comes to rapid revision of official public philosophy and policy the totalitarian nations obviously have an enormous advantage over the democracies.

control, and take care of the lower material needs; that this is the way of evolving nature and also, when it comes to rules for progress, works better than the inverse. One of the best refutations of Marxism is Marxism itself in that it was not Marx's actions in satisfying his material needs for subsistence that changed the world, but his philosophy, visionary ideas, and Communist ideology.

A value system that puts its ultimate good in the welfare of the "Party," and at the same time pointedly scorns reverence for nature, does little to help remedy most of our mounting global crisis conditions which today are the overriding concern. There is little in Marxist doctrine to help control overpopulation, curb pollution, conserve resources, preserve the environment, or protect endangered species. Nature in Marxist materialism is not something to be revered but almost the reverse, i.e., something to be battled and subjugated, transformed, mechanized, and exploited to satisfy man's (mainly material) needs (3). The forces of nature as interpreted by Marx in the materialist tradition are blind and unprincipled, not rich in quality, wonder, and beauty, harmoniously controlled with countless checks and balances, and full of creative strategies, constraints, and principles that have been time-tested for success in creating, preserving, and improving the quality of the biosphere, including the mind and spirituality of man.

In Marxism, not nature but technology and production power are idolized. Cathedrals for Marx are factories and skyscrapers, and the beautiful dream is to transform whole continents by industrial progress with "huge new populations springing up as if by magic" (61). The narrow focus on class conflict in an industrialized society also does nothing for the major ailments of the planet today and again is expressed in terms of the mechanistic determinism of the more material and elemental needs and components in man's makeup at the expense of the higher psychological needs and more idealistic components. The causal power of cognitive ideals that our new mentalism recognizes today, was, on principle dismissed. Where nothing is sacred and there is no higher meaning (beyond that of the "Party"), everything loses meaning.

It should be noted that most of the above differences are direct consequences of an acceptance of the materialist philosophy and have been heavily represented in practice in capitalist countries also, as well as in the communist world. Our quarrel here is not personal or political but ideological. The question at issue is what the world picture of science stands for in reference to political ideals. Any homocentric emphasis on the more material needs and aims of man, technology, industry, and production power, combined with a demeaning view of nature, whether Marxist, capitalist, or otheriwse, seems to represent the epitome of the worst forces that have caught up with us today to produce most of the adverse crisis conditions that threaten the future.

In Summary

In the context of today's mounting global problems and in the absence of population controls the relative long-term social benefits from advances in science and technology are diminished. At the same time the human value spin-offs from the mind-brain and other sciences are thrust into a strategic position of top concern because of their key role in the search for ultimate criteria for policy priorities and decision-making guidelines. Recent conceptual developments in the mind-brain sciences rejecting reductionism and mechanistic determinism on the one side, and dualism on the other, clear the way for a rational approach to the theory of values and to a natural fusion of science with ethics and religion. Science can be upheld as the best route to an increased understanding and rapport with the forces that made and move understanding and rapport with the forces that made, and move and control the universe and created man. The outlines for a global ethic emerge that would promote a reverent respect for nature and for the evolving quality of the biosphere and in which, the well being, further development and sanctity of the human psyche stand out as the foremost, but not sole, concern. Such values, if implemented, would set in motion the kind of social changes needed to lead us out of the vicious spirals of worsening world conditions. For our children's children, averting nuclear war won't help much if the population bomb and other global threats continue unchecked.

8

Addendum: Issue of Primacy and the Highest Good

In a recent book, *The Biological Origin of Human Values*, Pugh (58) offers a value theory which accords in most respects with the hierarchic theory presented here but which parts company in placing the main emphasis on genetic and innate origins. Pugh designates the innate biological values as primary, while the more cognitive, acquired values we deal with here primarily are referred to by Pugh as secondary. The innate basis of values and the fact that the higher, acquired values are developed upon and from the innate values, as stressed by Pugh, are both accepted in our own scheme, but these innate components are largely dispensed with in our view as constituting a basic denominator common to all human value systems. Our concern is with the variables and the conflicts in human values and how to select and resolve these for future benefit. Primary attention accordingly is centered on the more cognitive, rational, acquired values which Pugh calls secondary.

While in some respects this may appear to be merely a matter of emphasis, the difference from our standpoint is not trivial and calls for further comment. When he gives primacy to the more basic, more primitive or animalistic values built-in by evolution, Pugh—like Marxism, behaviorism, and sociobiology—overlooks or rejects the key principle of emergent determinism or downward control stressed throughout the present treatise, whereby the more highly evolved entities in a system may overpower the action of the more basic components out of which they are formed and evolved.

I take for granted that there is much of the biological and animalistic in human value systems. Of more importance in my view is the fact that the higher rational and uniquely human properties may transcend and gain control over the lower biological properties as, for example, when a person sets fire to himself in the public square, goes on a work or hunger strike, or otherwise conducts himself in the interests of some "higher cause." In contending that the so-called secondary acquired values cannot supersede the innate values, Pugh's theory and the present are in direct contradiction.

In my scheme, values are perceived to be organized in a complex of nested manifolds involving value hierarchies within hierarchies. Values at different levels may be in harmony, or in conflict. If in conflict, the higher may overpower the lower, or vice versa. However, it is a distinguishing characteristic of the civilized person or society that the more natural biological values founded in our evolutionary ancestry are increasingly superseded and kept under control by higher and more cognitive, acquired guidelines: religious, cultural, legal, rational, etc.

In these hierarchies, the concept of the highest good becomes particularly critical as a major determinant of the value structure. The highest good and what is held sacred standing at the top of the value hierarchy, set subsidiary values and condition the form of the "good life" throughout. Unlike the innate values, the concept of what is sacred varies with different peoples and faiths, and is subject to change by reason and learning. The values that count most are not at all inherently fixed nor are they absolute or immutable. It is in the power of the human brain to set its sights on new ideals above the innate drives and adjust its values accordingly. To a very large extent the concern for human values and efforts to seek corrective adjustments for an updated, more evolved, more valid and improved ethic that will work for today's world and the future can be concentrated around the single question of what ought to be held most sacred.

In the generation of values I have recognized two major categories of determining variables, "internal" and "external," that act as co-functions and each of which is subject to a range of alternatives, choices among which make for critical differences in the shaping of one's value hierarchy. Among the external factors I consistently favor the world picture and kind of reality verified by science, as opposed to those derived from other sources. Among the "inner" or "mental constraint" variables, the more highly evolved are favored over the less evolved, and this includes placing the more rational, cognitive determinants above the innate biological factors given primacy in Pugh's theory.

Each inner mental state tends to have a value framework of its own. In the brain of a person who has been starved for a week, things leading to food acquire special value. To one out of work or financially pressed, the means of personal subsistence become uppermost at the expense of less immediate concerns such as conservation, environmental protection, etc. In the majority of people most of the time immediate personal needs tend to dominate the value structure, and to suppress more transcendent values relating to the long-term good of the planet or of mankind as a whole. Only in the higher mental states that transcend immediate personal demands does one hope to find the kinds of value priorities needed today at national and international levels if we are to change the present course of global deterioration.

These higher value perspectives of the transcendent mind that separate men from animals and the civilized from the primitive can be made to exert control over the effects of values lower in the human hierarchy (Marx, Pugh, and sociobiology to the contrary). This can be effected in various ways, through legislation, religious dedication, even by just plain will power. It is a main theme of the present argument that we stand to benefit by an active attempt to bring into play the corrective influence of these more highly evolved rational values of the transcendant mind to supersede less evolved values which, along with those based in otherworldly beliefs, now dominate and currently are failing. To counteract and overcome the many natural, more immediate social pressures in which the world seems to become more and more hopelessly entrapped will require a higher vision of the most powerful kind.

For the first time in human history, global conditions have reached a stage that demands value perspectives which transcend not only the innate biological drives but even traditional humanitarian guidelines that have been respected for centuries. What may appear today to be most humane, compassionate and civically and morally upright, may later prove to be the most inhumane, cruel, and sinful when viewed from the standpoint of those many hundreds of generations hopefully to come. Even for the immediate good of this, our own generation, it now becomes important that new, long-term, more godlike guidelines—of a kind that will insure long survival and further progress in the quality of life—be instituted very soon if humanity is to live again with a sense of hope, purpose or higher meaning.

Notes

1. Values: Number One Problem of Our Times

This first effort to follow up some of the value implications of the changed view of consciousness was prepared originally as a lecture for a centennial symposium on "Biological Controls and Human Values" at Ohio State University, May 1970, that was abruptly canceled following the Kent State riots. The lecture was presented subsequently in the 1971 Honors Program on "Earth and Myth" at the University of Houston, under the title "Value and Belief in a Scientific World." The present version titled "Science and the Problem of Values" was accepted for publication in February 1972 by Dwight Ingle, editor of *Perspectives in Biology and Medicine* and friend and colleague of Ralph Burhoe, founder of *Zygon, Journal of Religion and Science*. Burhoe in particular had been sympathetic and quick to grasp the value implications of the changed views in mind-brain science. Later the same year Burhoe launched a rejuvenated push on values and proposed establishment in Chicago of the *Center for Advanced Studies in Religion and Science*. The present edited text is reprinted by permission of the University of Chicago Press from *Perspectives in Biology and Medicine*, volume 16 (1972).

The theme of this 1972 message has continued to hold up and grow stronger with time, receiving support also from others, among them Lester R. Brown, head of Worldwatch Institute, Washington, D.C., whose latest book, *Building a Sustainable Society*, includes a final chapter, "Changing Values and Shifting Priorities," which concurs that "values are the key to the evolution of a sustainable society not only because they influence behavior but also because they determine a society's priorities and thus its ability to survive." His timely and comprehensive treatment of the problem is strongly recommended.

2. Mind, Brain, and Humanist Values

Presented initially as a public "Monday Lecture" at the University of Chicago, May 1965, on request for a nonspecialized humanist discussion, this semi-popular lecture was published later in the same year by the University of Chicago Press in *New Views of the Nature of Man*, edited by John Platt. The present section on free-will is drawn from a later condensed version that was reprinted the following year in the *Bulletin of Atomic Scientists*. Although the essence of this new interpretation of consciousness had been expressed briefly in passing a year before in my James Arthur Lecture, this was the first extended publication and commitment to the mentalist position around which the remainder of the essays are developed. Although this mid-sixties essay represents the initial foundation and turning point toward the author's new philosophy, it is placed second here because the lead article more fully and directly presents the main theme of the volume.

3. The Ultimate Frame of Reference

This essay was presented originally as an address in Washington, D.C., November 1976, at the Fifth International Conference on the Unity of the Sciences sponsored by the International Cultural Foundation and chaired by Sir John Eccles.

4. Messages from the Laboratory

This chapter is taken from the main part of a response made on receiving the Passano Foundation Award for medical sciences at a dinner in Atlantic City, April 1973. I thank Edward Hutchings for valuable editorial suggestions incorporated on the occasion of its appearance in 1974 in Caltech's journal of *Engineering and Science*.

5. Bridging Science and Values

This chapter is adapted from a paper presented at a Neuroscience/ Philosophy Conference at the Claremont Colleges, California, February 1975. later versions were read at the International Conference on the Centrality of Science and Absolute Values, New York, December 1975, and at the meeting of the American Association for the Advancement of Science, Boston, February 1976. The present edited version from *American Psychologist*, volume 32 (1977), is reprinted by permission of the American Psychological Association.

6. Mind-Brain Interaction: Mentalism, Yes; Dualism, No

The chapter was written originally for a volume on mind-brain inter-action planned by D. L. Wilson, P. Glotzbach, and M. Ringle, focused around the interactionist theme of the book, *The Self and Its Brain*, by Karl Popper and John Eccles. The book was to have included a response by Popper and Eccles, which later had to be withdrawn, whereupon the projected volume was canceled. The article was later published in *Neu-roscience*, volume 5(1980), and is here reprinted by permission of the Society for Neuroscience.

7. Changing Priorities

This most recent of the collected writings was composed in early 1980 on invitation to do a prefatory chapter for the upcoming *Annual Reviews of Neuroscience*. It contains a summary of many of the high points of the other articles, and its contents overall come closest to being representative of the theme of the volume as a whole.

References

1. Armstrong, D. M. 1968. *A Materialist Theory of the Mind*. London: Routledge & Kegan Paul.
2. Bahm, A. J. 1974. *Ethics as a Behavioral Science*. Springfield, Ill.: Charles C. Thomas.
3. Bell, D. 1975. Technology, nature, and society. In *The Frontiers of Knowledge*. The Frank Nelson Doubleday Lectures. Garden City, N.Y.: Doubleday.
4. Beloff, J. 1962. *The Existence of Mind*. New York: Citadel Press.
5. Bindra, D. 1970. The problem of subjective experience: Puzzlement on reading R. W. Sperry's "A modified concept of consciousness." *Psychol. Rev.* 77:581–584.
6. Blanshard, B. and B. F. Skinner. 1967. The problem of consciousness—A debate. *Philosophy and Phenomenological Research* 27:317–337. Reprinted in M. H. Marx and F. E. Goodson, eds., *Theories in Contemporary Psychology*, pp. 205–223. New York: Macmillan, 1976.
7. Boring, E. G. 1942. *Sensation and Perception in the History of Experimental Psychology*. New York: Appleton-Century.
8. Brown, L. R. 1979. Consultation column. *InterDependent* 6:1–5.
9. Bunge, M. 1977. Emergence and the mind. *Neuroscience* 2:501–510.
10. Burhoe, R. W. 1969. Values via science. *Zygon* 4:65–99.
11. Burhoe, R. W. 1975. The human prospect and the "lord of history." *Zygon* 10:299–375.
12. Carnap, R. 1938. Logical Foundations of the Unity of Science. *Encyclopedia of Unified Science*, 1:42–62. Chicago: University of Chicago Press.
13. Cattell, R. B. 1972. *A New Morality from Science: Beyondism*. New York: Pergamon.
14. Dember, W. N. 1974. Motivation and the cognitive revolution. *Am. Psychol.* 29:161–168.
15. Dobzhansky, T. 1967. *Biology of Ultimate Concern*. New York: New American Library.
16. Doty, R. W. 1975. Consciousness from neurons. *Acta Neurobiologiae Experimentalis* 35:791–804.

17. Dubos, R. 1968. *So Human an Animal.* New York: Scribner's.
18. Eccles, J. C. 1953. *The Neurophysiological Basis of Mind: The Principles of Neurophysiology.* Oxford: Clarendon Press.
19. Eccles, J. C., ed. 1966. *Brain and Conscious Experience.* New York: Springer.
20. Eccles, J. C. 1968. The importance of brain research for the educational, cultural, and scientific future of mankind. *Perspect. Biol. Med.* 12:61–68.
21. Eccles, J. C. 1973. Brain, speech, and consciousness. *Die Naturwissenschaften* 60:167–176.
22. Eccles, J. C. 1973. *The Understanding of the Brain.* New York: McGraw-Hill.
23. Eccles, J. C. 1979. *The Human Mystery.* Gifford Lectures, 1978. Berlin: Springer.
24. Ehrlich, P. R. 1968. *The Population Bomb.* New York: Ballantine.
25. Eibel-Eibesfeldt, T. 1970. *Ethology: The Biology of Behavior.* New York: Holt, Rinehart and Winston.
26. Feigl, H. 1953. Unity of science and unitary science. In H. Feigl and M. Brodbeck, eds., *Readings in the Philosophy of Science,* pp. 382–384. New York: Appleton-Century Crofts.
27. Feigl, H. 1967. The "mental" and the "physical." In H. Feigl, M. Scriven, and G. Maxwell, eds., *Concepts, Theories, and the Mind-Body Problem.* Minneapolis: University of Minnesota Press.
28. Freeman, J. D. 1979. Towards an anthropology both scientific and humanistic. *Canberra Anthropology* 1:44–69.
29. Fuller, J. L. and W. R. Thompson. 1960. *Behavior Genetics.* New York: Wiley.
30. Globus, G. G. 1973. Consciousness and brain. I. The Identity Thesis. *Archs. Gen. Psychiat.* 29:153–160.
31. Gray, J. A. 1971. The mind-brain identity theory as a scientific hypothesis. *Philosoph. Q.* 21:247–254.
32. Hardin, G. 1972. *Exploring New Ethics for Survival.* New York: Viking.
33. John, E. R. 1976. A model for consciousness. In G. E. Schwartz and D. Shapiro, eds., *Consciousness and Self Regulation.* New York: Plenum Press.
34. Jones, W. T. 1965. *The Sciences and the Humanities.* Los Angeles: University of California Press.
35. Kantor, J. R. 1978. Cognition as events and as psychic constructions. *Psychol. Rec.* 28:329–342.
36. Kluckholm, K. 1959. The scientific study of values and contemporary civilization. *Proc. Am. Philosophical Society.* 102:469–376.
37. Koffka, K. 1935. *Principles of Gestalt Psychology.* New York: Harcourt, Brace.
38. Köhler, W. 1929. *Gestalt Psychology.* New York: Liverwright.

39. Köhler, W. 1961. The mind-body problem. In S. Hook, ed., *Dimensions of Mind*, pp. 15–32. New York: Collier Books.
40. Köhler, W. and R. Held. 1949. The cortical correlate of pattern vision. *Science* 110:414–419.
41. Libet, B. 1973. Electrical stimulation of cortex in human subjects and conscious sensory aspects. In A. Iggo, ed., *Handbook of Sensory Physiology*, vol. 2. New York: Springer.
42. MacKay, D. M. 1966. Cerebral organization and the conscious control of action. In J. C. Eccles, ed., *Brain and Conscious Experience*, pp. 312f, 422–44. Heidelberg: Springer.
43. MacKay, D. M. 1978. Selves and brains. *Neuroscience* 3:599–606.
44. MacKay, D. M. 1980. *Brains, Machines, and Persons*. London: Collins.
45. Matson, F. W. 1971. Humanistic theory: the third revolution in psychology. *The Humanist* (March/April). Reprinted in (eds. P. Zimbardo and C. Maslach), *Psychology for Our Times*, pp. 19–25. Glenview, Ill.: Scott Foresman, 1973.
46. Mishan, J. 1969. *The Costs of Economic Growth*. New York: Praeger.
47. Morgan, C. Lloyd. 1923. *Emergent Evolution*. New York: Holt.
48. Oppenheim, P. and H. Putnam. 1958. Unity of science as a working hypothesis. In H. Feigl, M. Scriven, and G. Maxwell, eds., *Minnesota Studies in the Philosophy of Science Concepts, Theories, and the Mind-Body Problem*, 2:3–36. Minneapolis: University of Minnesota Press.
49. Perry, J. R. 1978. Defenses for the mind-brain identity theory: causal differences. *Behav. Brain Sci.* 3:362.
50. Platt, J. 1969. What we must do. *Science* 166:1115–1121.
51. Polanyi, M. 1964. *Science, Faith, and Society*. Chicago: University of Chicago Press.
52. Pols, E. 1971. Power and agency. *International Philosophical Q.* 11:293–313.
53. Pols, E. 1975. *Meditation on a Prisoner*. Carbondale: Southern Illinois University Press.
54. Popper, K. 1962. *Conjectures and Refutations: The Growth of Scientific Knowledge*. New York-London: Basic Books.
55. Popper, K. 1972. *Objective Knowledge*. Oxford, London: Clarendon Press.
56. Popper, K. 1978. Natural selection and emergence of mind. *Dialectica* 32:339–355.
57. Popper, K. and J. C. Eccles. 1977. *The Self and Its Brain: An Argument for Interactionism*. New York: Springer International.
58. Pugh, G. E. 1977. *The Biological Origin of Human Values*. New York: Basic Books.
59. Pylyshyn, Z. W. 1973. What the mind's eye tells the mind's brain: a critique of mental imagery. *Psychol. Bull.* 80:1–24.
60. Roszak, T. 1973. *Where the Wasteland Ends: Politics and Transcendance in Postindustrial Society*. Garden City, N.Y.: Doubleday.

61. Ryazanoff, D., ed. 1963. *The Communist Manifesto of Karl Marx and Friedrich Engles.* New York: Russel & Russel.
62. Sawhill, J. C. 1979. The role of science in higher education. *Science* 206(4416):281.
63. Skinner, B. F. 1971. *Beyond Freedom and Dignity.* New York: Knopf.
64. Skinner, B. F. 1974. *About Behaviorism.* New York: Knopf.
65. Smart, J. J. C. 1978. Cortical localization and the mind-brain identity theory. *Behav. Brain Sci.* 3:365.
66. Snow, C. P. 1959. *The Two Cultures and the Scientific Revolution.* New York: Cambridge University Press.
67. Sperry, R. W. 1952. Neurology and the mind-brain problem. *Am. Sci.* 40:291–312.
68. Sperry, R. W. 1958. Physiological plasticity and brain circuit theory. In H. F. Harlow and C. N. Woolsey, *Biological and Biochemical Bases of Behavior*, pp. 401–424. Madison: University of Wisconsin Press.
69. Sperry, R. W. 1963. Chemoaffinity in the orderly growth of nerve fiber patterns and connections. *Proc. Natl. Acad. Sci.* 50:703.
70. Sperry, R. W. 1964. The great cerebral commissure. *Sci. Amer.* 210:42–52.
71. Sperry, R. W. 1964. Problems outstanding in the evolution of brain function. James Arthur Lecture. American Museum of Natural History, New York. Reprinted in R. Duncan and M. Weston-Smith, eds., *The Encyclopaedia of Ignorance*, pp. 423–433. New York: Pergamon.
72. Sperry, R. W. 1965. Embryogenesis of behavioral nerve nets. In R. L. Dehaan and H. Ursprung, eds., *Organogenesis*, 6:161–185. New York: Holt, Rinehart & Winston.
73. Sperry, R. W. 1965. Mind, brain, and humanist values. In J. R. Platt, ed., *New Views of the Nature of Man.* Chicago: University of Chicago Press. Condensed in *Bull. Atomic Scientists* (1966), 22:2–6.
74. Sperry, R. W. 1966. Brain bisection and mechanisms of consciousness. In J. C. Eccles, ed., *Brain and Conscious Experience*, pp. 298–313. New York: Springer.
75. Sperry, R. W. 1969a. Toward a theory of mind. *Proc. Natl. Acad. Sci.* 1:230–231.
76. Sperry, R. W. 1969b. A modified concept of consciousness. *Psychol. Rev.* 76:532–536.
77. Sperry, R. W. 1970a. Perception in the absence of the neocortical commissures. In *Perception and its Disorders*, 68:123–138. Assoc. for Research in Nervous and Mental Disease.
78. Sperry, R. W. 1970b. An objective approach to subjective experience. Further explanation of a hypothesis. *Psychol. Rev.* 77:585–590.
79. Sperry, R. W. 1971. How a developing brain gets itself properly wired for adaptive function. In E. Tobach, E. Shaw, and L. R. Aaronson,

eds., *The Biopsychology of Development*, pp. 27–44. New York: Academic Press.

80. Sperry, R. W. 1972. Science and the problem of values. *Perspect. Biol. Med.* 16:115–130. Reprinted in *Zygon* (1974), 9:7–21.

81. Sperry, R. W. 1976a. Mental phenomena as causal determinants in brain function. In G. Globus, G. Maxwell, and I. Savodnik, eds., *Consciousness and the Brain*. New York: Plenum. Reprinted in *Process Studies* (1976), 5:247–256.

82. Sperry, R. W. 1976b. Changing concepts of consciousness and free will. *Perspect. Biol. Med.* 20:9–19.

83. Sperry, R. W. 1976c. A unifying approach to mind and brain: Ten year perspective. In M. A. Corner and D. F. Swaab, eds., *Perspectives in Brain Research*, vol. 45. Amsterdam: Elsevier Scientific.

84. Sperry, R. W. 1977a. Bridging science and values: a unifying view of mind and brain. *Am. Psychol.* 32:237–245. Reprinted in *Zygon* (1979), 14:7–21.

85. Sperry, R. W. 1977b. Forebrain commissurotomy and conscious awareness. *J. Med. Philos.* 2:101–126.

86. Sperry, R. W. 1978. Mentalist monism: consciousness as a causal emergent of brain processes. *Behav. Brain Sci.* 3:367.

87. Sperry, R. W. 1980. Mind-brain interaction: Mentalism, Yes; Dualism, No. *Neuroscience* 5:195–206. Reprinted in A. D. Smith, R. Llinas, and P. D. Kostyuk, eds., *Commentaries in the Neurosciences* Oxford: Pergamon Press, 1980.

88. Wann, T. W., ed. 1965. *Behaviorism and Phenomenology: Contrasting Bases for Modern Psychology*. Chicago: University of Chicago Press.

89. Ward, M. F. 1978. The mind-brain issue unsimplified. *Behav. Brain Sci.* 3:368–369.

90. Wilson, D. L. 1976. On the nature of consciousness and of physical reality. *Perspect. Biol. Med.* 19:568–581.

Already Published in
CONVERGENCE

Convergence
by
Ruth Nanda Anshen

"There is no use trying," said Alice; "one *can't* believe impossible things."

"I dare say you haven't had much practice," said the Queen. "When I was your age, I always did it for half an hour a day. Why, sometimes I've believed as many as six impossible things before breakfast."

This commitment is an inherent part of human nature and an aspect of our creativity. Each advance of science brings increased comprehension and appreciation of the nature, meaning, and wonder of the creative forces that move the cosmos and created man. Such openness and confidence lead to faith in the reality of possibility and eventually to the following truth: "The mystery of the universe is its comprehensibility."

When Einstein uttered that challenging statement, he could have been speaking about our relationship with the universe. The old division of the Earth and the Cosmos into objective processes in space and time and mind in which they are mirrored is no longer a suitable starting point for understanding the universe, science, or ourselves. Science now begins to focus on the convergence of man and nature, on the framework which makes us, as living beings, dependent parts of nature and simultaneously makes nature the object of our thoughts and actions. Scientists can no longer confront the universe as objective observers. Science recognizes the participation of man with the universe. Speaking quantitatively, the universe is largely indifferent to what happens in man. Speaking qualitatively, nothing happens in man that does not have a bearing on the elements which constitute the universe. This gives cosmic significance to the person.

Nevertheless, all facts are not born free and equal. There exists a hierarchy of facts in relation to a hierarchy of values. To arrange the facts rightly, to differentiate the important from the trivial, to see their bearing in relation to each other and to evaluational criteria, requires a judgment which is intuitive as well as empirical. Man needs meaning in addition to information. Accuracy is not the same as truth.

Our hope is to overcome the cultural *hubris* in which we have been living. The scientific method, the technique of analyzing, explaining, and classifying has demonstrated its inherent limitations. They arise because, by its intervention, science presumes to alter and fashion the object of its investigation. In reality, method and object can no longer be separated. The outworn Cartesian, scientific world view has ceased to be scientific in the most profound sense of the word, for a common bond links us all— man, animal, plant, and galaxy—in the unitary principle of all reality. For the self without the universe is empty.

This universe of which we human beings are particles may be defined as a living, dynamic process of unfolding. It is a breathing universe, its respiration being only one of the many rhythms of its life. It is evolution itself. Although what we observe may seem to be a community of separate, independent units, in actuality these units are made up of subunits, each with a life of its own, and the subunits constitute smaller living entities. At no level in the hierarchy of nature is independence a reality. For that which lives and constitutes matter, whether organic or inorganic, is dependent on discrete entities that, gathered together, form aggregates of new units which interact in support of one another and become an unfolding event, in constant motion, with ever-increasing complexity and intricacy of their organization.

Are there goals in evolution? Or are there only discernible patterns? Certainly there is a law of evolution by which we can explain the emergence of forms capable of activities which are indeed novel. Examples may be said to be the origin of life, the emergence of individual consciousness, and the appearance of language.

The hope of the concerned authors in Convergence is that they will show that evolution and development are interchangeable and that the entire system of the interweaving of man, nature, and the universe constitutes a living totality. Man is searching for his legitimate place in this unity, this cosmic scheme of things. The meaning of this cosmic scheme—if indeed we can impose meaning on the mystery and majesty of nature—and the extent to which we can assume responsibility in it as uniquely intelligent beings, are supreme questions for which this Series seeks an answer.

Inevitably, toward the end of a historical period, when thought and custom have petrified into rigidity and when the elaborate machinery of civilization opposes and represses our more noble qualities, life stirs again beneath the hard surface. Nevertheless, this attempt to define the purpose of Convergence is set forth with profound trepidation. We are living in a period of extreme darkness. There is moral atrophy, destructive radiation within us, as we watch the collapse of values hitherto cherished—but now betrayed. We seem to be face to face with an apocalyptic destiny. The anomie, the chaos, surrounding us produces an almost lethal disintegration of the person, as well as ecological and demographic disaster. Our situation is desperate. And there is no glossing over the deep and unresolved tragedy that fills our lives. Science now begins to question its premises and tells us not only what *is*, but what *ought* to be, reconciling order and hierarchy.

My description of Convergence is not to be construed as an afterword essay for each individual volume. These few pages attempt to set forth the general aim and purpose of this Series. It is my hope that this statement will provide the reader with a new orientation in his thinking, one more specifically defined by these scholars who have been invited to participate in this intellectual, spiritual, and moral endeavor so desperately needed in our time. These scholars recognize the relevance of the nondiscursive experience of life which the discursive, analytical method alone is unable to convey.

The authors invited to Convergence Series acknowledge a structural kinship between subject and object, between living and nonliving matter, the immanence of the past energizing the present and thus bestowing a promise for the future. This kinship has long been sensed and experienced by mystics. Saint Francis of Assisi described with extraordinary beauty the truth that the more we know about nature, its unity with all life, the more we realize that we are one family, summoned to acknowledge the intimacy of our familial ties with the universe. At one time we were so anthropomorphic as to exclude as inferior such other aspects of our relatives as animals, plants, galaxies, or other species—even inorganic matter. This only exposed our provincialism. Then we believed there were borders beyond which we could not, must not, trespass. These frontiers have never existed. Now we are beginning to recognize, even take pride in, our neighbors in the Cosmos.

Human thought has been formed through centuries of man's consciousness, by perceptions and meanings that relate us to nature. The smallest living entity, be it a molecule or a particle, is at the same time present in the structure of the Earth and all its inhabitants, whether human or manifesting themselves in the multiplicity of other forms of life.

Today we are beginning to open ourselves to this evolved experience of consciousness. We keenly realize that man has intervened in the evolutionary process. The future is contingent, not completely prescribed, except for the immediate necessity to evaluate in order to live a life of integrity. The specific gravity of the burden of change has moved from genetic to cultural evolution. Genetic evolution itself has taken millions of years; cultural evolution is a child of no more than twenty or thirty thousand years. What will be the future of our evolutionary course? Will it be cyclical in the classical sense? Will it be linear in the modern sense? Yet we know that the laws of nature are not linear. Certainly, life is more than mere endless repetition. We must restore the importance of each moment, each deed. This is impossible if the future is nothing but a mechanical extrapolation of the past. Dignity becomes possible only with choice. The choice is ours.

In this light, evolution shows man arisen by a creative power inherent in the universe. The immense ancestral effort that has borne man invests him with a cosmic responsibility. Michelangelo's image of Adam created at God's command becomes a more intelligent symbol of man's position in the world than does a description of man as a chance aggregate of atoms or cells. Each

successive stage of emergence is more comprehensive, more meaningful, more fulfilling, and more converging, than the last. Yet a
higher faculty must always operate through the levels that are
below it. The higher faculty must enlist the laws controlling the
lower levels in the service of higher principles, and the lower level
which enables the higher one to operate through it will always
limit the scope of these operations, even menacing them with
possible failure. All our higher endeavors must work through our
lower forms and are necessarily exposed thereby to corruption.
We may thus recognize the cosmic roots of tragedy and our fallible
human condition. Language itself, as the power of universals, is
the basic expression of man's ability to transcend his environment
and to transmute tragedy into a moral and spiritual triumph.

This relationship, this convergence, of the higher with the lower
applies again when an upper level, such as consciousness or
freedom, endeavors to reach beyond itself. If no higher level can
be accounted for by the operation of a lower level, then no effort
of ours can be truly creative in the sense of establishing a higher
principle not intrinsic to our initial condition. And establishing
such a principle is what all great art, great thought, and great
action must aim at. This is indeed how these efforts have built up
the heritage in which our lives continue to grow.

Has man's intelligence broken through the limits of his own
powers? Yes and no. Inventive efforts can never fully account for
their success, but the story of man's evolution testifies to a creative
power that goes beyond that which we can account for in ourselves.
This power can make us surpass ourselves. We exercise some of
it in the simple art of acquiring knowledge and holding it to be
true. For, in doing so, we strive for intellectual control over things
outside ourselves, in spite of our manifest incapacity to justify this
hope. The greatest efforts of the human mind amount to no more
than this. All such acts impose an obligation to strive for the
ostensibly impossible, representing man's search for the fulfillment
of those ideals which, for the moment, seem to be beyond his
reach. For the good of a moral act is inherent in the act itself and
has the power to ennoble the person who performs it. Without
this moral ingredient there is corruption.

The origins of one person can be envisaged by tracing that

person's family tree all the way back to the primeval specks of protoplasm in which his first origins lie. The history of the family tree converges with everything that has contributed to the making of a human being. This segment of evolution is on a par with the history of a fertilized egg developing into a mature person, or the history of a plant growing from a seed; it includes everything that caused that person, or that plant, or that animal, or even that star in a galaxy, to come into existence. Natural selection plays no part in the evolution of a single human being. We do not include in the mechanism of growth the possible adversities which did not befall it and hence did not prevent it. The same principle of development holds for the evolution of a single human being; nothing is gained in understanding this evolution by considering the adverse chances which might have prevented it.

In our search for a reasonable cosmic view, we turn in the first place to common understanding. Science largely relies for its subject matter on a common knowledge of things. Concepts of life and death, plant and animal, health and sickness, youth and age, mind and body, machine and technical processes, and other innumerable and equally important things are commonly known. All these concepts apply to complex entities, whose reality is called into question by a theory of knowledge which claims that the entire universe should ultimately be represented in all its aspects by the physical laws governing the inanimate substrate of nature. "Technological inevitability" has alienated man's relationship with nature, with other human beings, with himself. Judgment, decision, and freedom of choice, in other words *knowledge* which contains a moral imperative, cannot be ordered in the form that some technological scientists believe. For there is no mechanical ordering, no exhaustive set of permutations or combinations that can perform the task. The power which man has achieved through technology has been transformed into spiritual and moral impotence. Without the insight into the nature of *being*, more important than *doing*, the soul of man is imperiled. And those self-transcendent ends that ultimately confer dignity, meaning, and identity on man and his life constitute the only final values worth pursuing. The pollution of consciousness is the result of mere technological efficiency. In addition, the authors in this Series recognize that the computer in itself can process information—not meaning. Thus we see on the stage of life no moral actors, only anonymous events.

Our new theory of knowledge, as the authors in this Series try to demonstrate, rejects this claim and restores our respect for the immense range of common knowledge acquired by our experience of convergence. Starting from here, we sketch out our cosmic perspective by exploring the wider implications of the fact that all knowledge is acquired and possessed by relationship, coalescence, convergence.

We identify a person's physiognomy by depending on our awareness of features that we are unable to specify, and this amounts to a convergence in the features of a person for the purpose of comprehending their joint meaning. We are also able to read in the features and behavior of a person the presence of moods, the gleam of intelligence, the response to animals or a sunset or a fugue by Bach, the signs of sanity, human responsibility, and experience. At a lower level, we comprehend by a similar mechanism the body of a person and understand the functions of the physiological mechanism. We know that even physical theories constitute in this way the processes of inanimate nature. Such are the various levels of knowledge acquired and possessed by the experience of convergence.

The authors in this Series grasp the truth that these levels form a hierarchy of comprehensive entities. Inorganic matter is comprehended by physical laws; the mechanism of physiology is built on these laws and enlists them in its service. Then, the intelligent behavior of a person relies on the healthy functions of the body and finally, moral responsibility relies on the faculties of intelligence directing moral acts.

We realize how the operations of machines, and of mechanisms in general, rely on the laws of physics but cannot be explained, or accounted for, by these laws. In a hierarchic sequence of comprehensive levels, each higher level is related to the levels below it in the same way as the operations of a machine are related to the particulars, obeying the laws of physics. We cannot explain the operations of an upper level in terms of the particulars on which its operations rely. Each higher level of integration represents, in this sense, a higher level of existence, not completely accountable by the levels below it yet including these lower levels implicitly.

In a hierarchic sequence of comprehensive levels each higher level is known to us by relying on our awareness of the particulars on the level below it. We are conscious of each level by internalizing its particulars and mentally performing the integration that constitutes it. This is how all experience, as well as all knowledge, is based on convergence, and this is how the consecutive stages of convergence form a continuous transition from the understanding of the inorganic, the inanimate, to the comprehension of man's moral responsibility and participation in the totality, the organismic whole, of all reality. The sciences of the subject-object relationship thus pass imperceptibly into the metascience of the convergence of the subject and object interrelationship, mutually altering each other. From the minimum of convergence, exercised in a physical observation, we move without a break to the maximum of convergence, which is a total commitment.

"The last of life, for which the first was made, is yet to come." Thus, Convergence has summoned the world's most concerned thinkers to rediscover the experience of *feeling*, as well as of thought. The convergence of all forms of reality presides over the possible fulfillment of self-awareness—not the isolated, alienated self, but rather the participation in the life process with other lives and other forms of life. Convergence is a cosmic force and may possess liberating powers allowing man to become what he is, capable of freedom, justice, love. Thus man experiences the meaning of grace.

A further aim of this Series is not, or could it be, to disparage science. The authors themselves are adequate witness to this fact. Actually, in viewing the role of science, one arrives at a much more modest judgment of its function in our whole body of knowledge. Original knowledge was probably not acquired by us in the active sense; most of it must have been given to us in the same mysterious way we received our consciousness. As to content and usefulness, scientific knowledge is an infinitesimal fraction of natural knowledge. Nevertheless, it is knowledge whose structure is endowed with beauty because its abstractions satisfy our urge for specific knowledge much more fully than does natural knowledge, and we are justly proud of scientific knowledge because we can call it our own creation. It teaches us clear thinking, and the extent to which clear thinking helps us to order our sensations is a marvel which fills the mind with ever new and increasing admiration and awe. Science now begins to include the realm of human values, lest even the memory of what it means to be human be forgotten.

Organization and energy are always with us, wherever we look, on all levels. At the level of the atom, organization becomes indistinguishable from form, from order, from whatever the forces are that hold the spinning groups of ultimate particles together in their apparent solidity. And now that we are at the atomic level, we find that modern physics has recognized that these ultimate particles are primarily electrical charges, and that mass is therefore a manifestation of energy. This has often been misinterpreted by idealists as meaning that matter has somehow been magicked away as if by a conjuror's wand. But nothing could be more untrue. It is impossible to transform matter into spirit just by making it thin. Bishop Berkeley's views admit of no refutation but carry no conviction nevertheless. However, something has happened to matter. It was only separated from form because it seemed too simple. Now we realize that, and this is a revolutionary change, we cannot separate them. We are now summoned to cease speaking of Form and Matter and begin to consider the convergence of Organization and Energy. For the largest molecule we know and the smallest living particles we know overlap. Such a cooperation, even though far down at the molecular level, cannot but remind us of the voluntary cooperation of the individual human beings in maintaining patterns of society at levels of organization far higher. The tasks of Energy and Organization in the making of the universe and ourselves are far from ended.

No individual destiny can be separated from the destiny of the universe. Alfred North Whitehead has stated that every event, every step or process in the universe, involves both effects from past situations and the anticipation of future potentialities. Basic for this doctrine is the assumption that the course of the universe results from a multiple and never-ending complex of steps developing out of one another. Thus, in spite of all evidence to the contrary, we conclude that there is a continuing and permanent energy of that which is not only man but all of life. For not an atom stirs in matter, organic and inorganic, that does not have its cunning duplicate in mind. And faith in the convergence of life with all its multiple manifestations creates its own verification.

We are concerned in this Series with the unitary structure of all nature. At the beginning, as we see in Hesiod's *Theogony* and in the Book of Genesis, there was a primal unity, a state of fusion in which, later, all elements become separated but then merge again. However, out of this unity there emerge, through separation, parts of opposite elements. These opposites intersect or reunite, in meteoric phenomena or in individual living things. Yet, in spite of the immense diversity of creation, a profound underlying convergence exists in all nature. And the principle of the conservation of energy simply signifies that there is a *something* that remains constant. Whatever fresh notions of the world may be given us by future experiments, we are certain beforehand that something remains unchanged which we may call *energy*. We now do not say that the law of nature springs from the invariability of God, but with that curious mixture of arrogance and humility which scientists have learned to put in place of theological terminology, we say instead that the law of conservation is the physical expression of the elements by which nature makes itself understood by us.

The universe is our home. There is no other universe than the universe of all life including the mind of man, the merging of life with life. Our consciousness is evolving, the primordial principle of the unfolding of that which is implied or contained in all matter and spirit. We ask: Will the central mystery of the cosmos, as well as man's awareness of and participation in it, be unveiled, although forever receding, asymptotically? Shall we perhaps be able to see all things, great and small, glittering with new light and reborn meaning, ancient but now again relevant in an iconic image which is related to our own time and experience?

The cosmic significance of this panorama is revealed when we consider it as the stages of an evolution that has achieved the rise of man and his consciousness. This is the new plateau on which we now stand. It may seem obvious that the succession of changes,

sustained through a thousand million years, which have transformed microscopic specks of protoplasm into the human race,
has brought forth, in so doing, a higher and altogether novel kind
of being, capable of compassion, wonder, beauty, and truth,
although each form is as precious, as sacred, as the other. The
interdependence of everything with everything else in the totality
of being includes a participation of nature in history and demands
a participation of the universe.

The future brings us nothing, gives us nothing; it is we who in
order to build it have to give it everything, our very life. But to be
able to give, one has to possess; and we possess no other life, no
living sap, than the treasures stored up from the past and digested,
assimilated, and created afresh by us. Like all human activities,
the law of growth, of evolution, of convergence draws its vigor
from a tradition which does not die.

At this point, however, we must remember that the law of
growth, of evolution, has both a creative and a tragic nature. This
we recognize as a degenerative process, as devolution. Whether it
is the growth of a human soul or the growth of a living cell or of
the universe, we are confronted not only with fulfillment but with
sacrifice, with increase and decrease, with enrichment and dimi-
nution. Choice and decision are necessary for growth and each
choice, each decision, excludes certain potentialities, certain potential realities. But since these unactualized realities are part of
us, they possess a right and command of their own. They must
avenge themselves for their exclusion from existence. They may
perish and with them all the potential powers of their existence,
their creativity. Or they may not perish but remain unquickened
within us, repressed, lurking, ominous, swift to invade in some
disguised form our life process, not as a dynamic, creative,
converging power, but as a necrotic, pathological force. If the
diminishing and the predatory processes co-mingle, atrophy and
even death in every category of life ensue. But if we possess the
maturity and the wisdom to accept the necessity of choice, of
decision, of order and hierarchy, the inalienable right of freedom
and autonomy, then, in spite of its tragedy, its exclusiveness, the
law of growth endows us with greatness and a new moral dimen-
sion.

Convergence is committed to the search for the deeper meanings of science, philosophy, law, morality, history, technology, in fact all the disciplines in a transdisciplinary frame of reference. This Series aims to expose the error in that form of science which creates an unreconcilable dichotomy between the observer and the participant, thereby destroying the uniqueness of each discipline by neutralizing it. For in the end we would know everything but *understand nothing*, not being motivated by concern for any question. This Series further aims to examine relentlessly the ultimate premises on which work in the respective fields of knowledge rest and to break through from these into the universal principles which are the very basis of all specialist information. More concretely, there are issues which wait to be examined in relation to, for example, the philosophical and moral meanings of the models of modern physics, the question of the purely physico-chemical processes versus the postulate of the irreducibility of life in biology. For there is a basic correlation of elements in nature, of which man is a part, which cannot be separated, which compose each other, which converge, and alter each other mutually.

Certain mysteries are now known to us the mystery, in part, of the universe and the mystery of the mind have been in a sense revealed out of the heart of darkness. Mind and matter, mind and brain, have converged; space, time, and motion are reconciled; man, consciousness, and the universe are reunited since the atom in a star is the same as the atom in man. We are homeward bound because we have accepted our convergence with the Cosmos. We have reconciled observer and participant. For at last we know that time and space are modes by which we think, but not conditions in which we live and have our being. Religion and science meld; reason and feeling merge in mutual respect for each other, nourishing each other, deepening, quickening, and enriching our experiences of the life process. We have heeded the haunting voice in the Whirlwind.

About the Author

Roger Sperry, neurobiologist and pioneer brain researcher with early roots in psychology and the humanities, holds the Hixon Chair as Professor of Psychobiology at the California Institute of Technology. In 1981 he shared the Nobel prize in Medicine/Physiology for providing "an insight into the inner world of the brain which hitherto had been almost completely hidden from us."

Sperry won early acclaim for discovering how the brain is able to inherit and grow selectively its own networks for behavior unaided by function. Later with his students he innovated the "split-brain" studies that brought an upgraded view of the right half of the brain and opened a large new field of investigation in the mental specialties of the two sides of the brain. His concept of consciousness as causal, introduced in the mid-sixties, helped bring an end to the reign of behaviorism in Psychology.

Recipient of many awards in medicine and science, he was cited by the American Psychological Association for "fundamental contributions to our knowledge of the nature of man" and by the National Society for Neuroscience for contributing "radical conceptual conversions in two different fields of inquiry" along with "an entirely novel theory of consciousness."

He is a member of the Pontifical Academy of Sciences, a Fellow of the Royal Society of Great Britain, and the American Philosophical Society, among others, and has been awarded honorary degrees from Cambridge University in England, the Rockefeller University, the University of Chicago, Kenyon College and his alma mater, Oberlin College. Dr. Sperry's honors include the William Thompson Wakeman and Passano Foundation Awards in

Medicine, and the Howard Crosby Warren Medal in Experimental Psychology. In 1979 he received the Wolf Foundation Prize in Medicine presented at the Knesset in Jerusalem, the Ralph Gerard Prize, and the Albert Lasker Award, the highest prize in American medicine.